GRANULAR COMPUTING

ANALYSIS AND DESIGN OF INTELLIGENT SYSTEMS

INDUSTRIAL ELECTRONICS SERIES

Series Editors:

Bogdan M. Wilamowski & J. David Irwin

PUBLISHED TITLES

Granular Computing: Analysis and Design of Intelligent Systems, *Witold Pedrycz*

Industrial Wireless Sensor Networks: Applications, Protocols, Standards, and Products,
V. Çağrı Güngör and Gerhard P. Hancke

Power Electronics and Control Techniques for Maximum Energy Harvesting in
Photovoltaic Systems, *Nicola Femia, Giovanni Petrone, Giovanni Spagnuolo,
and Massimo Vitelli*

Extreme Environment Electronics, *John D. Cressler and H. Alan Mantooth*

Renewable Energy Systems: Advanced Conversion Technologies and Applications,
Fang Lin Luo and Hong Ye

Multiobjective Optimization Methodology: A Jumping Gene Approach,
K.S. Tang, T.M. Chan, R.J. Yin, and K.F. Man

The Industrial Information Technology Handbook, *Richard Zurawski*

The Power Electronics Handbook, *Timothy L. Skvarenina*

Supervised and Unsupervised Pattern Recognition: Feature Extraction and
Computational Intelligence, *Evangelia Micheli-Tzanakou*

Switched Reluctance Motor Drives: Modeling, Simulation, Analysis, Design,
and Applications, *R. Krishnan*

FORTHCOMING TITLES

Smart Grid Technologies: Applications, Architectures, Protocols, and Standards,
*Vehbi Cagri Gungor, Carlo Cecati, Gerhard P. Hancke, Concettina Buccella,
and Pierluigi Siano*

Multilevel Converters for Industrial Applications, *Sergio Alberto Gonzalez,
Santiago Andres Verne, and Maria Ines Valla*

Data Mining: Theory and Practice, *Milos Manic*

Electric Multiphase Motor Drives: Modeling and Control, *Emil Levi, Martin Jones,
and Drazen Dujic*

Sensorless Control Systems for AC Machines: A Multiscalar Model-Based Approach,
Zbigniew Krzeminski

Next-Generation Optical Networks: QoS for Industry, *Janusz Korniak and Pawel Rozycki*

Signal Integrity in Digital Systems: Principles and Practice, *Jianjian Song and
Edward Wheeler*

FPGAs: Fundamentals, Advanced Features, and Applications in Industrial Electronics,
Juan Jose Rodriguez Andina and Eduardo de la Torre

Dynamics of Electrical Machines: Practical Examples in Energy and Transportation
Systems, *M. Kemal Saioglu, Bulent Bilir, Metin Gokasan, and Seta Bogosyan*

GRANULAR COMPUTING

ANALYSIS AND DESIGN OF INTELLIGENT SYSTEMS

Witold Pedrycz

CRC Press
Taylor & Francis Group
Boca Raton London New York

CRC Press is an imprint of the
Taylor & Francis Group, an **informa** business

CRC Press
Taylor & Francis Group
6000 Broken Sound Parkway NW, Suite 300
Boca Raton, FL 33487-2742

First issued in paperback 2017

ISBN-13: 978-1-4398-8681-6 (hbk)
ISBN-13: 978-1-138-07449-1 (pbk)

Library of Congress Cataloging-in-Publication Data

Pedrycz, Witold, 1953-
    Granular computing : analysis and design of intelligent systems / author, Witold Pedrycz.
    pages cm. -- (Industrial electronics series)
    Summary: "Given the nature of the technology, granular computing cuts across a broad range of engineering disciplines. This self-contained book builds upon introductory ideas and provides illustrative examples that help facilitate a better grasp of more advanced material and enhance its overall presentation. It will be of a particular appeal to those engaged in research and practical developments in computer, electrical, industrial, manufacturing, and biomedical engineering. It will be equally well suited for those coming from nontechnical disciplines where information granules assume a highly visible position."-- Provided by publisher.
    Includes bibliographical references and index.
    ISBN 978-1-4398-8681-6 (hardback)
    1. Granular computing. I. Title.

QA76.9.S63P4423 2013
006.3--dc23
                                       2013007328

Visit the Taylor & Francis Web site at
http://www.taylorandfrancis.com

and the CRC Press Web site at
http://www.crcpress.com

To Ewa, Barbara, and Adam

Contents

Preface

Undoubtedly, intelligent systems come with various connotations, exhibit different semantics, and imply numerous realizations. In spite of the existing diversity of views and opinions, there are several features that are quite profoundly visible across a broad spectrum of those systems. We anticipate that to some extent an intelligent system should be able to effectively communicate with the user, seamlessly accept requests from the human, and convey the results in a transparent, meaningful, and easy to comprehend format. A user-friendly, two-way communication between the human and the system arises as an imperative feature of intelligent systems. Furthermore, an intelligent system has to be endowed with substantial learning capabilities where learning can be carried out not only in the presence of numeric data but also in situations when the system is exposed to nonnumeric evidence, perceptions, opinions, and judgments. These forms of interaction and learning/adaptation are difficult to realize at the level of plain numbers but have to be carried out at the level of more abstract, carefully structured entities—information granules. There are piles of data everywhere—in engineering, marketing, banking, Internet, insurance, chemical plants, and transportation networks. We face a genuine challenge: how to make sense of them, to help the user understand what they tell, and to gather useful evidence to support processes of decision making. It is apparent that these processes have to lead to constructs that form a useful abstraction of original data. In response to these growing challenges, Granular Computing (GrC) emerged a decade ago as a unified conceptual and processing framework.

Information granules, as encountered in natural language, are *implicit* in their nature. To make them fully operational so that they become effectively used in the analysis and design of intelligent systems, we need to make information granules *explicit*. This is possible through a prudent formalization available within the realm of GrC. Among the existing possibilities offered by GrC, we may refer to sets (interval calculus), fuzzy sets, rough sets, shadowed sets, probabilistic granules, just to name some of the well-established alternatives. Each of them comes with its own methodological setup, comprehensive design framework, and a large body of knowledge supporting analysis, design, and processing of constructs developed therein.

There is an important, visible, and timely tendency in the formation of information granules. We strive to build them in a way that they are in rapport with the implicit character of information granules. Some representative examples of this tendency arise in the form of higher order, higher type, and hybrid information granules. For instance, we can talk about fuzzy sets of type 2 or order 2, rough fuzzy sets, linguistic probabilities, and so forth.

Z-numbers just coined a year ago, come as yet another important and highly
visible construct reflective of this timely tendency.

Granular Computing embraces an extensive body of knowledge, which dwells
upon the individual formalisms of information granules and unifies them to
form a coherent methodological and developmental environment. As a discipline, GrC is still *in statu nascendi* where fundamentals, concepts, and design
methodologies are very much needed. One of the ultimate objectives of this book
is to discuss several principles of GrC. The principle of justifiable granularity
supports ways of forming information granules based on numeric and granular
(perception-based) experimental evidence. Information granularity is viewed as
a crucial design asset that helps endow existing modeling paradigms and modeling practice with new conceptual and algorithmic features, thus making the
resulting models more reflective of the complexities of real-world phenomena.
The optimization of information granularity is essential in facilitating processes
of collaboration and consensus seeking, in which a significant deal of flexibility is required. In turn, this flexibility is effectively utilized through a prudent
distribution of information granularity across individual models involved in
consensus building. The granulation–degranulation principle captures a way of
representing any datum with the use of a collection of information granules and
quantifying the associated representation error.

The volume is self-contained to a significant degree. While some general and
quite limited preliminaries of set theory, fuzzy sets, and system modeling could
be found beneficial, the exposure of the material builds upon some introductory
ideas and comes with illustrative examples that help facilitate a better grasp of
more advanced material and enhance its overall presentation. Given the nature
of the technology, GrC cuts across a broad range of engineering disciplines. It
will be of particular appeal to those engaged in research and practical developments in computer, electrical, industrial, manufacturing, and biomedical
engineering. It will be equally well suited for those coming from nontechnical
disciplines where information granules assume a highly visible position.

The organization of the material reflects the main objectives of this book.
The sequence of chapters is formed in the following way:

Chapter 1 offers a comprehensive introduction to the concepts of information granules, information granularity, and Granular Computing, and raises
a number of arguments and applications. Several fundamentals of Granular
Computing are highlighted.

Chapter 2 presents the key formalisms of information granules and elabo-
rates on the underlying processing supported by each of them.

Chapter 3 builds on the concepts of information granules and their formalisms by augmenting existing representations through an introduction of
higher-order and higher-type information granules.

Chapter 4 discusses an operational concept of information granulation
and degranulation by highlighting the essence of this tandem and its quantification in terms of the associated error.

Chapters 5 and 6 are focused on the two principles of GrC. The first one, the principle of justifiable granularity, supports a design of information granules regarded as a process in which one aims at the formation of a compromise between justifiability and specificity of information granules to be constructed. The second principle stresses the need to look at the information granularity as an important design asset, which helps construct more realistic models of real-world systems or facilitate collaborative pursuits of system modeling.

Chapter 8 follows the main theme discussed in Chapter 6 by venturing in more detail into granular models by highlighting their concepts, architectures, and design algorithms.

Chapters 7, 9, 10, and 11 elaborate on the selected application domains in which GrC and granular models play a visible role. These areas include pattern recognition, time series, and decision making.

Depending upon the interest of the reader, several possible routes are envisioned in the figure below.

As shown in the diagram, in most cases there is a linear flow of exposure of the material. The most introductory line of presentation that might be useful to those interested in having a quick view of the principles of GrC would include a sequence of Chapters 1, 2, 4, and 5 with an eventual inclusion of

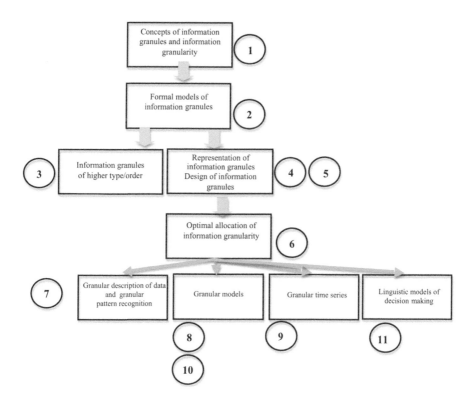

Chapter 3. The main track composed of Chapters 1, 2, 3, 4, and 5 covers the principles and the underlying methodology as well as the algorithmic backbone of GrC. Depending on the interest of the reader, this track could be augmented by Chapters 7 and 8; 9 and 10; and 11, either individually or as several combinations of them.

Additional information is available from the Taylor & Francis/CRC Web site: http://www.crcpress.com/product/isbn/9781439886816.

Witold Pedrycz
Department of Electrical and Computer Engineering
University of Alberta
Edmonton, Canada

The Author

Witold Pedrycz, Ph.D., is Professor and Canada Research Chair (CRC) in Computational Intelligence in the Department of Electrical and Computer Engineering, University of Alberta, Edmonton, Canada. He is also with the Systems Research Institute of the Polish Academy of Sciences, Warsaw, Poland and King Abudalaziz University, Saudi Arabia. In 2009, Dr. Pedrycz was elected a foreign member of the Polish Academy of Sciences. He is a Fellow of the Royal Society of Canada and a Fellow of the Institute of Electronic and Electrical Engineers (IEEE), International Fuzzy Systems Association (IFSA), International Society of Management Engineers, Engineers Canada, and The Engineering Institute of Canada.

Dr. Pedrycz's main research directions involve computational intelligence, fuzzy modeling and Granular Computing, knowledge discovery and data mining, fuzzy control, pattern recognition, knowledge-based neural networks, relational computing, and software engineering. He has published numerous papers in this area. He is also the author of 15 research monographs covering various aspects of computational intelligence and software engineering. Dr. Pedrycz has been a member of numerous program committees of IEEE conferences in the area of fuzzy sets and neurocomputing.

Dr. Pedrycz is intensively involved in editorial activities. He is Editor-in-Chief of *Information Sciences* and Editor-in-Chief of *IEEE Transactions on Systems, Man, and Cybernetics, Part A*. He currently serves as an Associate Editor of *IEEE Transactions on Fuzzy Systems* and a number of other international journals. In 2007, he received the prestigious Norbert Wiener award from the IEEE Systems, Man, and Cybernetics Society. Dr. Pedrycz is a recipient of the IEEE Canada Computer Engineering Medal. In 2009, he received a Cajastur Prize for Soft Computing from the European Centre for Soft Computing for "pioneering and multifaceted contributions to Granular Computing."

Foreword

Dr. Pedrycz's latest magnum opus, *Granular Computing Analysis and Design of Intelligent Systems*, or IS-GrC for short, breaks new ground in many directions. IS-GrC takes an important step toward achievement of human-level machine intelligence—a principal goal of artificial intelligence (AI) since its inception. In recent years, AI has achieved important successes, but achievement of human-level machine intelligence is still a fairly distant objective. There is a reason.

Humans have many remarkable capabilities. Among them there are two that stand out in importance. First, is the capability to graduate perceptions, that is, associate degrees with perceptions. More concretely, humans can easily answer questions exemplified by: On the scale from 0 to 10, how pleased are you with your job? On the scale from 0 to 10, how supportive are you of the President? The ability to graduate perceptions plays a key role in elicitation of membership functions of fuzzy sets.

The second key capability is that of granulation of perceptions. Informally, a granule is a clump of values of a perception (e.g., perception of age), which are drawn together by proximity, similarity, or functionality. More concretely, a granule may be interpreted as a restriction on the values that a variable can take. In this sense, words in a natural language are, in large measure, labels of granules. A linguistic variable is a variable whose values are words or, equivalently, granules. For example, *age* is a linguistic variable when its values are represented as very young, young, middle-aged, old, very old, and so forth. The concept of a linguistic variable was introduced in my 1973 paper "Outline of a new approach to the analysis of complex systems and decision processes." The concept of granulation is rooted in the concept of a linguistic variable. My first paper on information granularity was my 1979 paper, "Fuzzy Sets and Information Granularity."[*] Today, almost all applications of fuzzy logic involve the concept of a linguistic variable. What is of historical interest is that my 1973 paper was met with skepticism and derision, largely because numbers are respected but words are not.

The concept of granulation plays an essential role in human cognition, especially in the realm of everyday reasoning. Graduation and granulation have a position of centrality in fuzzy logic. In essence, fuzzy logic is the logic of classes with unsharp boundaries. What is striking is that neither graduation nor granulation, have been considered in AI. What is widely unrecognized is that without the machinery of graduation and granulation, it is not possible to achieve human-level machine intelligence. This is the principal

---

[*] Zadeh, L.A. 1979, Fuzzy Sets and Information Granularity, in: M. Gupta, R. Ragade, and R. Yager (Eds.), *Advances in Fuzzy Set Theory and Applications*, North-Holland, Amsterdam, 3–18, 1–32.

reason why achievement of human-level machine intelligence remains a distant goal in AI.

In Granular Computing (GrC), graduation and granulation play principal roles. Basically, in GrC the objects of reasoning and computation are not values of variables but granules of values of variables or, equivalently, restrictions on values of variables. This is a key distinction between traditional computing and Granular Computing. The principal distinction between GrC and computing with words (CWW) is that in CWW semantics of natural language plays a more important role than in GrC.

In coming years, GrC is certain to grow in visibility and importance. What will become increasingly apparent is that much of the information on which decisions are based is imperfect, that is, imprecise, uncertain, incomplete, and unreliable. Furthermore, much of the information is described in natural language. To deal with such information, what is needed are the machineries of GrC and CWW. Viewed against this backdrop, Dr. Pedrycz's magnum opus is a remarkably well put together and reader-friendly collection of concepts and techniques, which constitute Granular Computing.

Granular Computing is based on fuzzy logic. To make IS-GrC largely self-contained and reader-friendly, IS-GrC includes a succinct exposition of those concepts and formalisms within fuzzy logic, which are of relevance to Granular Computing. In addition, the introductory chapters focus on some of the basic concepts, which relate to information granularity, among them the concept of shadow sets and fuzzy sets of higher type and higher order. Dr. Pedrycz is a highly skilled expositor. His skill manifests itself in the introductory chapters and throughout the book. Each chapter ends with conclusions and references.

To refer to IS-GrC as a magnum opus is an understatement. IS-GrC combines extraordinary breadth with extraordinary depth. It contains a wealth of new ideas, and unfolds a vast panorama of concepts, methods, and applications.

A theme that is developed at length in IS-GrC relates to the rationale for granulation. Words in a natural language may be viewed as labels of granules. In science, there is a deep-seated tradition of according much more respect to numbers than to words. Indeed, progression from the use of words to the use of numbers is viewed as a hallmark of scientific progress. Seen in this perspective, use of words in GrC and, more generally in fuzzy logic, may appear to be a step in the wrong direction. In fact, this is not the case. In my writings, I refer to the use of words in place of numbers as the Fuzzy Logic Gambit. Underlying the Fuzzy Logic Gambit, there are two principal rationales. First, the use of words is a necessity when numbers are not known. Second, the use of words is an advantage when there is a tolerance for imprecision which can be exploited through the use of words. These two rationales play a key role in GrC.

The body of IS-GrC contains a large number of new concepts and new techniques. Particularly worthy of note are techniques which relate to classification and pattern recognition; granular mappings and allocation;

fuzzy rough sets and rough fuzzy sets; granular neural networks; granular description of data; construction of granular representations; architectures of granular models; structure-free granular models; granular time series; fuzzy logic networks and granular fuzzy rule-based systems, including the Takagi–Sugeno system; linguistic models of decision making, including the analytic hierarchy process (AHP); granulation of objective functions; and, models of processing with granular characterization of fuzzy sets.

What should be noted is that there is an issue in GrC, which is not easily resolved. The issue relates to informativeness. By its nature, granular information is imprecise. When the objects of computation are information granules, the result of computation will, in general, be less precise than the arguments. In some cases, underlying computation with information granules is the worst-case analysis. A challenging problem is that of developing a mode of analysis other than the worst-case analysis. This is a particularly important issue in the realm of decisions, which are based on granular information.

Dr. Pedrycz's development and description of these concepts, techniques, and their applications is a truly remarkable achievement. IS-GrC is a must reading for all who are concerned with the design and application of intelligent systems. Dr. Pedrycz and the publisher deserve a loud applause.

Lotfi A. Zadeh
University of California, Berkeley

1

Information Granularity, Information Granules, and Granular Computing

In this chapter, we briefly introduce the concept of information granules and processing of information granules. Information granules along with a faculty of their processing constitute an area of Granular Computing. We stress an omnipresent visibility of information granules in the perception of real-world phenomena, coping with their complexity, and highlight a role of information granules in user-friendly schemes of interpretation of processing results and the design of intelligent systems. Implicit and explicit facets of information granularity are noted. We elaborate on a variety of formal methods used to describe information granules and build the underlying processing framework. A bird's-eye view of the associated formalisms and ensuing semantics of information granularity and information granules is presented.

1.1 Information Granularity and the Discipline of Granular Computing

Information granules are intuitively appealing constructs, which play a pivotal role in human cognitive and decision-making activities. We perceive complex phenomena by organizing existing knowledge along with available experimental evidence and structuring them in a form of some meaningful, semantically sound entities, which are central to all ensuing processes of describing the world, reasoning about the environment, and supporting decision-making activities.

The term *information granularity* itself has emerged in different contexts and numerous areas of application. It carries various meanings. One can refer to artificial intelligence in which case information granularity is central to a way of problem solving through problem decomposition where various subtasks could be formed and solved individually. Information granules and the area of intelligent computing revolving around them being termed *Granular Computing* are quite often presented with a direct association with the pioneering studies by Zadeh (1997, 1999, 2005). Zadeh coined an informal yet highly descriptive notion of an information granule. In a general sense, by information granule, one regards a collection of elements drawn together by their closeness (resemblance, proximity, functionality, etc.,) articulated in terms of some useful spatial, temporal, or functional relationships.

Subsequently, Granular Computing is about representing, constructing, and processing information granules.

It is again worth stressing that information granules permeate almost all human endeavors (Bargiela and Pedrycz, 2003, 2005a,b, 2009; Zadeh, 1997, 1999; Pedrycz and Bargiela, 2002; Pedrycz, 2005a,c). No matter which problem is taken into consideration, we usually set it up in a certain conceptual framework composed of some generic and conceptually meaningful entities—information granules, which we regard to be of relevance to the problem formulation, further problem solving, and a way in which the findings are communicated to the community. Information granules realize a framework in which we formulate generic concepts by adopting a certain level of abstraction. Let us refer here to some areas, which offer compelling evidence as to the nature of underlying processing and interpretation in which information granules play a pivotal role:

Image processing. In spite of the continuous progress in the area, a human being assumes a dominant and very much uncontested position when it comes to understanding and interpreting images. Surely, we do not focus our attention on individual pixels and process them as such but group them together into semantically meaningful constructs—familiar objects we deal with in everyday life. Such objects involve regions that consist of pixels or categories of pixels drawn together because of their proximity in the image, similar texture, color, and so forth. This remarkable and unchallenged ability of humans dwells on our effortless ability to construct information granules, manipulate them, and arrive at sound conclusions.

Processing and interpretation of time series. From our perspective we can describe them in a semiqualitative manner by pointing at specific regions of such signals. Medical specialists can effortlessly interpret various diagnostic signals including electrocardiography (ECG) or electroencephalography (EEG) recordings. They distinguish some segments of such signals and interpret their combinations. In the stock market, one analyzes numerous time series by looking at amplitudes, trends, and patterns. Experts can interpret temporal readings of sensors and assess a status of the monitored system. Again, in all these situations, the individual samples of the signals are not the focal point of the analysis, synthesis, and the signal interpretation. We always granulate all phenomena (no matter if they are originally discrete or analog in their nature). When working with time series, information granulation occurs in time and in the feature space where the data are described.

Granulation of time. Time is another important and omnipresent variable that is subjected to granulation. We use seconds, minutes, days, months, and years. Depending upon a specific problem we have in mind and who the user is, the size of information granules (time intervals) could vary quite significantly. To high-level management, time intervals of quarters of a year or a few years could be meaningful temporal information granules on the basis of which one develops any predictive model. For those in charge of everyday operation of a dispatching center, minutes and hours could form a viable scale of time granulation. Long-term planning is very much different from day-to-day operation. For the designer of high-speed integrated circuits and digital systems, the temporal information granules concern nanoseconds, microseconds, and perhaps microseconds. Granularity of information (in this case time) helps us focus on the most suitable level of detail.

Design of software systems. We develop software artifacts by admitting a modular structure of an overall architecture of the designed system where each module is a result of identifying essential functional closeness of some components of the overall system. Modularity (granularity) is a holy grail of the systematic software design supporting a production of high-quality software products.

Information granules are examples of abstractions. As such they naturally give rise to hierarchical structures: the same problem or system can be perceived at different levels of specificity (detail) depending on the complexity of the problem, available computing resources, and particular needs to be addressed. A hierarchy of information granules is inherently visible in processing of information granules. The level of detail (which is represented in terms of the size of information granules) becomes an essential facet facilitating a way of hierarchical processing of information with different levels of hierarchy indexed by the size of information granules.

Even such commonly encountered and simple examples presented above are convincing enough to lead us to ascertain that (a) information granules are the key components of knowledge representation and processing, (b) the level of granularity of information granules (their size, to be more descriptive) becomes crucial to the problem description and an overall strategy of problem solving, (c) hierarchy of information granules supports an important aspect of perception of phenomena and deliver a tangible way of dealing with complexity by focusing on the most essential facets of the problem, and (d) there is no universal level of granularity of information; commonly the size of granules is problem oriented and user dependent.

Human centricity comes as an inherent feature of intelligent systems. It is anticipated that a two-way effective human–machine communication is imperative. Humans perceive the world, reason, and communicate at some level of abstraction. Abstraction comes hand in hand with nonnumeric constructs, which embrace collections of entities characterized by some notions of closeness, proximity, resemblance, or similarity. These collections are referred to as information granules. Processing of information granules is a fundamental way in which people process such entities. Granular Computing has emerged as a framework in which information granules are represented and manipulated by intelligent systems. The two-way communication of such intelligent systems with the users becomes substantially facilitated because of the usage of information granules.

By no means is the above quite descriptive definition of information granules formal. It rather intends to emphasize the crux of the idea and link it to human centricity and computing with perceptions rather than plain numbers (Pedrycz, 2005a,c; Pedrycz and Gacek, 2002; Bargiela and Pedrycz, 2008).

What has been said so far touches upon a qualitative aspect of the problem. The visible challenge is to develop a computing framework within which all these representation and processing endeavors could be formally realized.

While the notions of information granularity and information granules themselves are convincing, they are not operational (algorithmically sound) until some formal models of information granules along with the related algorithmic framework have been introduced. In other words, to secure the algorithmic realization of Granular Computing, the *implicit* nature of information granules has to be translated into the constructs that are *explicit* in their nature, that is, described formally in which information granules can be efficiently computed with.

The common platform emerging within this context comes under the name of Granular Computing. In essence, it is an emerging paradigm of information processing. While we have already noticed a number of important conceptual and computational constructs built in the domain of system modeling, machine learning, image processing, pattern recognition, and data compression in which various abstractions (and ensuing information granules) came into existence, Granular Computing becomes an innovative and intellectually proactive endeavor that manifests in several fundamental ways:

- It identifies the essential commonalities between the surprisingly diversified problems and technologies used there, which could be cast into a unified framework known as a granular world. This is a fully operational processing entity that interacts with the external world (that could be another granular or numeric world) by collecting necessary granular information and returning the outcomes of the Granular Computing.

- With the emergence of the unified framework of granular processing, we get a better grasp as to the role of interaction between various formalisms and visualize a way in which they communicate.

- It brings together the existing plethora of formalisms of set theory (interval analysis) (Moore, 1966), fuzzy sets (Zadeh, 1965, 2005), rough sets (Pawlak, 1982, 1985, 1991; Pawlak and Skowron, 2007a,b) under the same roof by clearly visualizing that in spite of their visibly distinct underpinnings (and ensuing processing), they exhibit some fundamental commonalities. In this sense, Granular Computing establishes a stimulating environment of synergy between the individual approaches.

- By building upon the commonalities of the existing formal approaches, Granular Computing helps assemble heterogeneous and multifaceted models of processing of information granules by clearly recognizing the orthogonal nature of some of the existing and well established frameworks (such as probability theory coming with its probability density functions and fuzzy sets with their membership functions).

- Granular Computing fully acknowledges a notion of variable granularity, whose range could cover detailed numeric entities and very abstract and general information granules. It looks at the aspects of compatibility of such information granules and ensuing communication mechanisms of the granular worlds.

- Granular Computing gives rise to processing that is less time demanding than the one required when dealing with detailed numeric processing (Bargiela and Pedrycz, 2003, 2005a,b).
- Interestingly, the inception of information granules is highly motivated. We do not form information granules without reason. Information granules arise as an evident realization of the fundamental paradigm of abstraction.

On the one hand, Granular Computing as an emerging area brings a great deal of original, unique ideas. On the other, it dwells substantially on the existing well-established developments that have already happened in a number of individual areas. In a synergistic fashion, Granular Computing brings fundamental ideas of interval analysis, fuzzy sets, and rough sets, and facilitates building a unified view of them where an overarching concept is the granularity of information itself. It helps identify main problems of processing and its key features, which are common to all the formalisms being considered.

Granular Computing forms a unified conceptual and computing platform. Yet, what is important is that it directly benefits from the already existing and well-established concepts of information granules formed in the setting of set theory, fuzzy sets, rough sets, and others. Reciprocally, the general investigations carried out under the rubric of Granular Computing offer some interesting and stimulating thoughts to be looked at within the realm of the specific formalism of sets, fuzzy sets, shadowed sets, or rough sets.

1.2 Formal Platforms of Information Granularity

There is a plethora of formal platforms in which information granules are defined and processed.

Sets (intervals) realize a concept of abstraction by introducing a notion of dichotomy: we admit an element to belong to a given information granule or to be excluded from it. Along with set theory comes a well-developed discipline of interval analysis (Moore, 1966). Alternatively to an enumeration of elements belonging to a given set, sets are described by characteristic functions taking on values in $\{0,1\}$. A characteristic function describing set A is defined as follows

$$A(x) = \begin{cases} 1, & \text{if } x \in A \\ 0, & \text{if } x \notin A \end{cases} \tag{1.1}$$

where A(x) stands for a value of the characteristic function of set A at point x. With the emergence of digital technologies, interval mathematics has appeared as an important discipline encompassing many applications. A family of sets defined in a universe of discourse **X** is denoted by $P(\mathbf{X})$. Well-known set operations—union, intersection, and complement are the three fundamental constructs supporting a manipulation on sets. In terms of the characteristic functions, they result in the following expressions,

$$(A \cap B)(x) = \min(A(x),B(x)) \ (A \cup B)(x) = \max(A(x),B(x)) \ \bar{A}(x) = 1 - A(x) \quad (1.2)$$

where A(x) and B(x) are the values of the characteristic functions of A and B at x, and $\bar{A}$ denotes the complement of A.

Fuzzy sets (Zadeh, 1965; Pedrycz and Gomide, 1998, 2007) provide an important conceptual and algorithmic generalization of sets. By admitting partial membership of an element to a given information granule we bring an important feature which makes the concept be in rapport with reality. It helps working with the notions where the principle of dichotomy is neither justified nor advantageous. The description of fuzzy sets is realized in terms of membership functions taking on values in the unit interval. Formally, a fuzzy set *A* is described by a membership function mapping the elements of a universe X to the unit interval [0,1] (Zadeh, 1965):

$$A: \mathbf{X} \rightarrow [0,1] \quad\quad\quad (1.3)$$

The membership functions are therefore synonymous with fuzzy sets. In a nutshell, membership functions generalize characteristic functions in the same way as fuzzy sets generalize sets. A family of fuzzy sets defined in **X** is denoted by $F(\mathbf{X})$. Fuzzy sets are generalizations of sets and are represented as a family of nested sets (representation theorem).

Operations on fuzzy sets are realized in the same way as already shown by Equation (1.2). However, given the fact that we are concerned with membership grades in [0,1], there are a number of alternatives in the realization of fuzzy sets operators (logic *and* and *or* operators, respectively). Those are implemented through so-called t-norms and t-conorms (Klement, Mesiar, and Pap, 2000; Pedrycz and Gomide, 1998, 2007).

Shadowed sets (Pedrycz, 1998, 1999, 2005b,c) offer an interesting description of information granules by distinguishing among elements, which fully belong to the concept, which are excluded from it, and whose belongingness is completely *unknown*. Formally, these information granules are described as a mapping X: $\mathbf{X} \rightarrow \{1, 0, [0,1]\}$ where the elements with the membership quantified as the entire [0,1] interval are used to describe a shadow of the construct. Given the nature of the mapping here, shadowed sets can be sought as a granular description of fuzzy sets where the shadow is used to localize unknown membership values, which in fuzzy sets are distributed over the entire universe of discourse. Note that the shadow produces nonnumeric descriptors of membership grades. A family of fuzzy sets defined in **X** is denoted by $S(\mathbf{X})$.

Probability-oriented information granules are expressed in the form of some probability density functions or probability functions. They capture a collection of elements resulting from some experiment. In virtue of the concept of probability, the granularity of information becomes a manifestation of the occurrence of some elements. For instance, each element of a set comes with a probability density function truncated to [0,1], which quantifies a degree of membership to the information granule. There are a number of variations of these constructs with probabilistic sets (Hirota, 1981; Hirota and Pedrycz, 1984) being one of them.

Rough sets (Pawlak, 1982, 1985, 1991; Pawlak and Skowron, 2007a,b; Lin and Cercone, 1997) emphasize a roughness of description of a given concept X when being realized in terms of the indiscernability relation provided in advance. The roughness of the description of X is manifested in terms of its lower and upper approximations of a certain rough set. A family of fuzzy sets defined in **X** is denoted by $R(\mathbf{X})$.

Figure 1.1 highlights the main differences between selected sets, fuzzy sets, rough sets, and shadowed sets. As noted, the key aspect is about a representation of a notion of belongingness of elements to the concept. It is binary in case of sets (yes-no, dichotomy), partial membership (fuzzy sets), lower and upper bound (rough sets), and uncertainty region (shadowed sets). Rough sets and shadowed sets, even though they exhibit conceptual differences, give rise to a construct that highlights some items endowed with uncertainty (undefined belongingness).

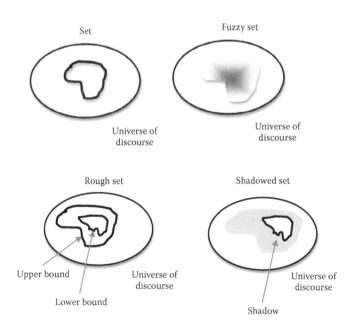

FIGURE 1.1
Conceptual realization of information granules—a comparative view.

Other formal models of information granules involve axiomatic sets (Liu and Pedrycz, 2009), soft sets, and intuitionistic sets (which are considered notions of membership and nonmembership, regarded as two independent concepts).

The choice of a certain formal setting of information granulation is mainly dictated by the formulation of the problem and the associated specifications coming with the problem. There is an interesting and quite broad spectrum of views on information granules and their processing. The two extremes are quite visible here:

Symbolic perspective. A concept—information granule is viewed as a single symbol (entity). This view is very much present in the artificial intelligence (AI) community, where computing revolves around symbolic processing. Symbols are subject to processing rules giving rise to results, which are again symbols coming from the same vocabulary one has started with.

Numeric perspective. Here information granules are associated with a detailed numeric characterization. Fuzzy sets are profound examples in this regard. We start with numeric membership functions. All ensuing processing involves *numeric* membership grades so in essence it focuses on number crunching. The results are inherently numeric. The progress present here has resulted in a diversity of numeric constructs. Because of the commonly encountered numeric treatment of fuzzy sets, the same applies to logic operators (connectives) encountered in fuzzy sets.

There are a number of alternatives of describing information that are positioned in-between these two extremes or the descriptions could be made more in a multilevel fashion. For instance, one could have an information granule described by a fuzzy set whose membership grades are symbolic (ordered terms such as *small, medium, high*; all defined in the unit interval).

With regard to the formal settings of information granules as briefly highlighted above, it is instructive to mention that all of them offer some operational realization (in different ways, though) of implicit concepts by endowing them with a well-defined semantics. For instance, treated implicitly an information granule *small* error is regarded just as a symbol (and could be subject to symbolic processing as usually realized in artificial intelligence), however once explicitly articulated as an information granule, it becomes associated with some semantics (becomes calibrated) thus coming with a sound operational description such as characteristic or membership function. In the formalism of fuzzy sets, the symbol *small* comes with the membership description, which could be further processed.

1.3 Information Granularity and Its Quantification

In a descriptive way, a level of abstraction supported by information granules is associated with the number of elements embraced by the granule. A certain measure of cardinality, which counts the number of elements in the information granule forms a sound descriptor of information granularity. The higher the granularity (the number of elements embraced by the information granule), the higher the abstraction of this granule and the lower its specificity. Starting with a set A, its cardinality is computed in the form of the sum (finite universe of discourse)

$$\text{card}(A) = \sum_{i=1}^{n} A(x_i) \tag{1.4}$$

or an integral (when the universe is infinite)

$$\text{card}(A) = \int_X A(x)\,dx \tag{1.5}$$

where A(x) is a formal description of the information granule (such as characteristic or membership function). For a fuzzy set, we count the number of its elements but one has to bear in mind that each element may belong with a certain degree of membership so the calculations carried out above involve the degrees of membership. In this case, one commonly refers to Equations (1.4) and (1.5) as a σ-count of A. For rough sets, one can proceed in a similar way as above by expressing the granularity of the lower and upper bounds of the rough set (roughness). In the case of probabilistic information granules, one can consider its standard deviation as a descriptor of granularity.

1.4 Information Granules and a Principle of the Least Commitment

Results of computing completed in the setting of Granular Computing come in the form of information granules. The specificity (level of detail) of such information granule if not sufficient (high) enough, may stress the necessity of refining already collected experimental evidence. Instead, committing a hustle decision made on the basis of perhaps insufficient information, the low level of specificity signals postponement of any action until more evidence has been collected and the obtained results become specific enough. The underlying idea is the one conveyed by a so-called principle of the least commitment. As an example, let us consider a decision-making problem in which several alternatives are sought and one of them needs to be picked up. If the levels of identified preferences come with these alternatives, that is, the corresponding membership degrees, they serve as a clear indicator quantifying hesitation as to the ability to make a decision. In other words, in light of the collected evidence, we do not intend to commit ourselves to making any decision (selection of one of the alternatives) as of yet. The intent would be to postpone decision and collect more compelling evidence. This could involve further collecting of data, soliciting expert opinion, and alike. Based on this new evidence, we could continue with computing eventually arriving at a resulting information granule of sufficiently high specificity. This stresses that information granularity delivers a computationally sound setup to realize the principle of the least commitment (Figure 1.2).

1.5 Information Granules of Higher Type and Higher Order

The formal concepts of information granules presented so far can be generalized. Two main directions of such generalizations are considered resulting in information granules of higher type and higher order. The first category is about the generalization that concerns ways of generalizing grades of membership (belongingness). The second one is about the increased level of sophistication of the universe of discourse in which information granules are constructed with the elements of this universe being a collection of information granules themselves.

Higher-type information granules. The quantification of levels of belongingness to a given information granule is granular itself rather than numeric as encountered in sets or fuzzy sets. This type of quantification is of interest in

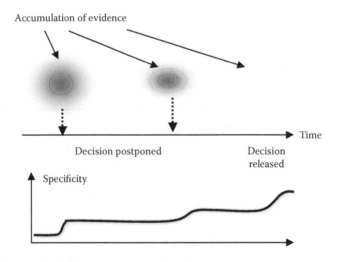

FIGURE 1.2
An essence of the principle of the least commitment; the decision is postponed until the moment the results become specific enough. Shown is a change (increase) in the levels of specificity occurring over time.

situations where it is not quite justifiable or technically sound to quantify the grade of membership in terms of a single numeric value. These situations give rise to ideas of type-2 fuzzy sets or interval-valued fuzzy sets. In the first case, the membership is quantified by a certain fuzzy set taking on values in the unit interval. In the second case, we have a subinterval of [0,1] representing membership values. One can discuss fuzzy sets of higher type in which the granular quantification is moved to the higher levels of the construct. For instance, one can talk about type-3, type-4,...fuzzy sets. Albeit conceptually sound, one should be aware that the computing overhead associated with further processing of such information granules becomes more significant. With this regard, type-2 fuzzy sets are of relevance here. In light of the essence of these constructs, we can view probabilistic granules treated as higher-type information granules as we admit membership values to be granulated in a probabilistic manner.

Higher-order information granules. The notion of higher order of information granules points at a space in which an information granule is defined. Here, the universe of discourse is composed of a family of information granules. For instance, a fuzzy set of order 2 is constructed in the space of a family of so-called reference fuzzy sets. This stands in sharp contrast with fuzzy sets of order 1, which are defined in individual elements of the universe of discourse. One could remark that fuzzy modeling quite often involves order-2 fuzzy sets.

The visualization underlying these concepts is included in Figure 1.3.

These types of construct could be generalized by invoking a number of consecutive levels of the structure thus giving rise to a hierarchy of information granules. In all situations, we have to assess whether moving to the higher-level or higher-order constructs is legitimate from the perspective of the problem at hand and serves the purpose of arriving at a more efficient and computationally appealing solution.

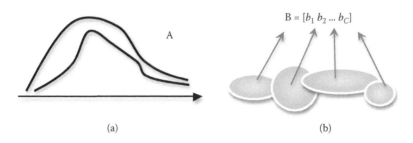

(a)   (b)

FIGURE 1.3
Examples of information granules of type-2 (type-2 fuzzy set A) (a) and order-2 fuzzy set B (defined over a collection of fuzzy sets) (b).

1.6 Hybrid Models of Information Granules

Information granules can embrace several granulation formalisms at the same time forming some hybrid models. These constructs become of particular interest when information granules have to capture a multifaceted nature of the problem. There are a large number of interesting options here. Some of them, which have been found convincing and practically relevant, include:

(a) *Fuzzy probabilities*. Probability and fuzzy sets are *orthogonal* concepts and as such they could be considered together as a single entity. The concepts of a fuzzy event and fuzzy probabilities (i.e., probabilities whose values are quantified in terms of fuzzy sets such as *high* probability, *very low* probability, *negligible* probability, etc.,) are of interest here.

(b) *Fuzzy rough and rough fuzzy information granules*. Here the indiscernibility relation can be formed on the basis of fuzzy sets. Fuzzy sets, rather than sets are also the entities that are described in terms of the elements of the indiscernibility relation. The original object X for which a rough set is formed might be a fuzzy set itself rather than a set used in the original definition of rough sets.

1.7 A Design of Information Granules

Information granules are human-centric constructs capturing the semantics of the abstract entities of interest in the setting of the problem at hand. They are directly reflective of the perception of concepts. Quite commonly they are constructed on the basis of some experimental data $\mathbf{D}$ and involve a certain granulation mechanism G. In general, we distinguish between the two main development scenarios:

(a) The data $\mathbf{D}$ give rise to a single information granule Ω,

$$G: \mathbf{D} \rightarrow \Omega \tag{1.6}$$

In this case, a general design scheme uses a principle of justifiable granularity (studied in Chapter 5). Its generality comes with the fact that it applies to any formalism of information granularity.

(b) The granulation results in a family of information granules A_1, A_2, ..., A_c,

$$G: \mathbf{D} \rightarrow \{A_1, A_2, ..., A_c\} \tag{1.7}$$

where usually we articulate some constraints on the family of the resulting granules. In both cases, the granulation mechanism involves a criterion of closeness of elements, and if required, could also embrace some aspects of functional resemblance. Clustering algorithms and their generalizations incorporating knowledge-enhanced clustering are commonly used (to be studied in Chapter 5).

1.8 The Granulation–Degranulation Principle

When dealing with any data point (object, pattern, etc.,) we are concerned with its representation in terms of some available collection of information granules. The granulation process forming a description of some object $\mathbf{x}$ in the form of information granules can be described as a mapping of the form where the result is a collection of membership grades (in the case of fuzzy sets),

$$G: \mathbf{R}^n \rightarrow [0,1]^c \tag{1.8}$$

In the sequel, the degranulation process realizes a reconstruction of $\mathbf{x}$ based on the collection of the grades of membership. The degranulation realizes a mapping of the form

$$G^{-1}: [0,1]^c \rightarrow \mathbf{R}^n \tag{1.9}$$

The capabilities of the information granules to reflect the structure of the original data can be conveniently expressed by comparing how much the result of degranulation, that is, $\hat{\mathbf{x}}$ differs from the original $\mathbf{x}$, that is, $\hat{\mathbf{x}} \neq \mathbf{x}$. Ideally, we may formally require that the following relationship holds: $\hat{\mathbf{x}} = G^{-1}(G(\mathbf{x}))$ where G and G^{-1} denote the corresponding phases of information granulation and degranulation. The algorithms of realizing the granulation–degranulation process are aimed at the minimization of a granulation error. Typically, the optimization of the granulation–degranulation is carried out in the presence of some data $\mathbf{K} = \{\mathbf{x}_1, \mathbf{x}_2, ..., \mathbf{x}_N\}$. The quality of the scheme (that is, a granulation–degranulation error) is expressed in the form of the sum of distances $\sum_{k=1}^{N} ||G^{--1}(G(\mathbf{x}_k)) - \mathbf{x}_k||$.

The granulation–degranulation process can be realized in the case of problems involving fuzzy sets of higher type in which a type reduction (from type-2 fuzzy sets to type-1 fuzzy sets) takes place.

1.9 Information Granularity in Data Representation and Processing

In data (signals, patterns) characterization, before proceeding with its detailed processing, a careful, prudently thought-out representation is an essential prerequisite directly impacting the effectiveness of all ensuing algorithms, in particular, influencing classification quality of pattern classifiers built on the basis of such information granules.

Granularity of information plays a primordial role in all these characterization and representation pursuits. Numeric data are represented in the form of information granules and the manifestation of such granules (e.g., as the values of membership functions) is used afterward in a more detailed system design.

Let us highlight the main role information granules play in the constructs of signal processing and system modeling. This role is visible in the formation of data interfaces. The essence of the underlying construct and its role vis-à-vis processing realized by any algorithm of signal processing is profoundly visible in the realization of a new feature space based on information granules. A general scheme is portrayed in Figure 1.4.

Formally speaking, the original data space, typically, n-dimensional space of real number vectors, $\mathbf{R}^n$, is transformed via a finite collection of information granules, that is, A_1, A_2, ..., A_c. We say that the input space has been granulated. Each input $\mathbf{x}$ is perceived by the following classifier/analyzer through the "eyes" of the information granules, resulting in the granulation process as captured by Equation (1.8). Recall that the result of the mapping is a c-dimensional vector positioned in the [0,1] hypercube. There are at least

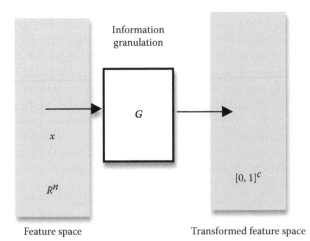

FIGURE 1.4
Formation of a new feature space through a collection of information granules.

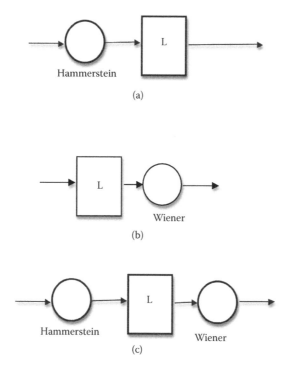

FIGURE 1.5
Architectures of Hammerstein, Wiener, and Hammerstein–Wiener classes of models. Shown are nonlinear interfaces of the linear module (L).

three important and practically advantageous aspects of the mapping realized by information granules:

Nonlinear mapping of the data space with an intent of forming information granules in such a way that the transformed data $G(x_1)$, $G(x_2)$, ..., $G(x_N)$ are more suitable to construct an effective classifier.

In the context of the general scheme presented in Figure 1.4, where the nonlinearity is delivered by information granules, it is worth recalling two general classes of models known as Hammerstein models and Wiener models as well as their combination known as Hammerstein–Wiener models (Pintelon and Schoukens, 2012). In all these cases, the linear component of the overall architecture is augmented by the nonlinear module in which inputs or outputs are transformed in a nonlinear way (see Figure 1.5).

Reduction of the dimensionality of the feature space. While the original feature space could be quite high (which is common in many classification problems), the dimensionality of the space of information granules is far lower, $c \ll n$. This supports the developments of the classifiers, especially neural networks and reduces a risk of memorization resulting in poor generalization capabilities. We often witness this role of information granules in the construction of neurofuzzy systems.

Information granules as essential constructs supporting the development of interpretable models. For instance, in rule-based systems (classifiers, analyzers), the condition parts (as well as conclusions) comprise information granules—interpretable entities, which make the rules meaningful.

1.10 Optimal Allocation of Information Granularity

Information granularity is an important design asset. Information granularity allocated to the original numeric model elevates a level of abstraction, that is, generalizes the original construct initially developed at the numeric level. A way in which such an asset is going to be distributed throughout the construct or a collection of constructs to make the abstraction more efficient, is a subject for optimization.

Consider a certain mapping $y = f(\mathbf{x}, \mathbf{a})$ with $\mathbf{a}$ being a vector of parameters of the mapping. The mapping can be sought in a general way. One may think of a fuzzy model, neural network, polynomial, differential equation, linear regression, and so forth. The granulation mechanism G is applied to $\mathbf{a}$ giving rise to its granular counterpart, $\mathbf{A} = G(\mathbf{a})$ and subsequently producing a granular mapping, $Y = G(f(\mathbf{x},\mathbf{a})) = f(\mathbf{x}, G(\mathbf{a})) = f(\mathbf{x}, \mathbf{A})$. Given the diversity of the underlying constructs as well as a variety of ways information granules can be formalized, we arrive at a suite of interesting constructs such as granular neural networks, that is, interval neural networks, fuzzy neural networks, probabilistic neural networks, and so forth.

There are a number of well-justified and convincing arguments behind elevating the level of abstraction of the existing constructs. These include: an ability to realize various mechanisms of collaboration, quantification of variability of sources of knowledge considered, and better modeling rapport with systems when dealing with nonstationary environments. These problems will be studied in depth in the following chapters.

1.11 Conclusions

This chapter offers a global view of the discipline of Granular Computing as a conceptual and algorithmic framework. Information granules are found in an array of applications once their implicit format is translated into the explicit (operational) format coming as a result of the expression of information granules in one of the available formal setups of sets, fuzzy sets, probabilities, and so forth. Some design strategies of information granules are elaborated on, including the principle of justifiable granularity.

References

Bargiela, A. and W. Pedrycz. 2003. *Granular Computing: An Introduction*. Dordrecht: Kluwer Academic Publishers.

Bargiela, A. and W. Pedrycz. 2005a. Granular mappings. *IEEE Transactions on Systems, Man, and Cybernetics*-Part A, 35, 2, 292–297.

Bargiela, A. and W. Pedrycz. 2005b. A model of granular data: A design problem with the Tchebyschev FCM. *Soft Computing*, 9, 155–163.

Bargiela, A. and W. Pedrycz. 2008. Toward a theory of Granular Computing for human-centered information processing. *IEEE Transactions on Fuzzy Systems*, 16, 2, 320–330.

Bargiela, A. and W. Pedrycz, eds. 2009. *Human-Centric Information Processing Through Granular Modelling*. Heidelberg: Springer-Verlag.

Hirota, K. 1981. Concepts of probabilistic sets. *Fuzzy Sets and Systems*, 5, 1, 31–46.

Hirota, K. and W. Pedrycz. 1984. Characterization of fuzzy clustering algorithms in terms of entropy of probabilistic sets. *Pattern Recognition Letters*, 2, 4, 213–216.

Klement, E., R. Mesiar, and E. Pap. 2000. *Triangular Norms*. Dordrecht: Kluwer Academic Publishers.

Lin, T.Y. and N. Cercone, eds. 1997. *Rough Sets and Data Mining*. Norwell, MA: Kluwer Academic Publishers.

Liu, X. and W. Pedrycz. 2009. *Axiomatic Fuzzy Set Theory and Its Applications*. Berlin: Springer-Verlag.

Moore, R. 1966. *Interval analysis*. Englewood Cliffs, NJ: Prentice-Hall.

Pawlak, Z. 1982. Rough sets. *Int. J. Comput. Inform. Sci.* 11, 341–356.

Pawlak, Z. 1985. Rough sets and fuzzy sets. *Fuzzy Sets and Systems*, 17, 1, 99–102.

Pawlak, Z. 1991. *Rough Sets. Theoretical Aspects of Reasoning About Data*. Dordrecht: Kluwer Academic Publishers.

Pawlak, Z. and A. Skowron. 2007a. Rudiments of rough sets. *Information Sciences*, 177, 1, 1, 3–27.

Pawlak, Z. and A. Skowron. 2007b. Rough sets and Boolean reasoning. *Information Sciences*, 177, 1, 41–73.

Pedrycz, W. 1998. Shadowed sets: Representing and processing fuzzy sets. *IEEE Trans. on Systems, Man, and Cybernetics, Part B*, 1, 28, 103–109.

Pedrycz, W. 1999. Shadowed sets: Bridging fuzzy and rough sets. In: *Rough Fuzzy Hybridization. A New Trend in Decision-Making*, S.K. Pal and A. Skowron (eds.). Singapore: Springer Verlag, 179–199.

Pedrycz, W. 2005a. From granular computing to computational intelligence and human-centric systems. *IEEE Connections*, 3, 2, 6–11.

Pedrycz, W. 2005b. Interpretation of clusters in the framework of shadowed sets. *Pattern Recognition Letters*, 26, 15, 2439–2449.

Pedrycz, W. 2005c. *Knowledge-Based Clustering: From Data to Information Granules*. Hoboken, NJ: John Wiley.

Pedrycz, W. and A. Bargiela. 2002. Granular clustering: A granular signature of data. *IEEE Trans. Systems, Man and Cybernetics*, 32, 212–224.

Pedrycz, W. and A. Gacek. 2002. Temporal granulation and its application to signal analysis. *Information Sciences*, 143, 1–4, 47–71.

Pedrycz, W. and F. Gomide. 1998. *An Introduction to Fuzzy Sets: Analysis and Design.* Cambridge, MA: MIT Press.

Pedrycz, W. and F. Gomide. 2007. *Fuzzy Systems Engineering: Toward Human-Centric Computing.* Hoboken, NJ: John Wiley.

Pintelon, R. and J. Schoukens. 2012. *System Identification: A Frequency Domain Approach,* 2nd ed. New York: J. Wiley-IEEE Press.

Zadeh, L.A. 1965. Fuzzy sets. *Information & Control,* 8, 338–353.

Zadeh, L.A. 1997. Towards a theory of fuzzy information granulation and its centrality in human reasoning and fuzzy logic. *Fuzzy Sets and Systems,* 90, 111–117.

Zadeh, L.A. 1999. From computing with numbers to computing with words—From manipulation of measurements to manipulation of perceptions. *IEEE Trans. on Circuits and Systems,* 45, 105–119.

Zadeh, L.A. 2005. Toward a generalized theory of uncertainty (GTU)—an outline. *Information Sciences,* 172, 1–40.

2

Key Formalisms for Representation of Information Granules and Processing Mechanisms

Information granules are formalized in many different ways. We review a number of highly representative formal frameworks of information granulation, emphasize their main features, show linkages among them, and highlight the main differences existing there. The essential computing machinery for information granules is presented. It is shown how some commonalities with this regard arise.

2.1 Sets and Interval Analysis

Sets and set theory are the fundamental notions of mathematics and science. They are in common usage when describing a wealth of concepts, describing relationships, and formalizing solutions. The underlying fundamental notion of set theory is that of *dichotomy*: a certain element belongs to a set or is excluded from it. A universe of discourse **X** over which a set or sets are formed could be very diversified depending upon the nature of the problem.

Given a certain element in a universe of discourse **X**, a process of dichotomization (binarization) imposes a binary, *all-or-none* classification decision: we either accept or reject the element as belonging to a given set. For instance, consider the set S shown in Figure 2.1. The point x_1 belongs to S whereas x_2 does not, that is, $x_1 \in S$ and $x_2 \notin S$. Similarly, for set T we have $x_1 \notin T$, and $x_2 \in T$.

If we denote the acceptance decision about the belongingness of the element by 1 and the reject decision (nonbelongingness) by 0, we express the classification (assignment) decision of $x \in \mathbf{X}$ to the given set (S or T) through a characteristic function:

$$S(x) = \begin{cases} 1, & \text{if } x \in S \\ 0, & \text{if } x \notin S \end{cases} \qquad T(x) = \begin{cases} 1, & \text{if } x \in T \\ 0, & \text{if } x \notin T \end{cases} \tag{2.1}$$

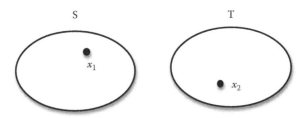

FIGURE 2.1
Examples of sets S and T.

In general, a characteristic function of set A defined in **X** assumes the following form

$$A(x) = \begin{cases} 1, \text{ if } x \in A \\ 0, \text{ if } x \notin A \end{cases} \qquad (2.2)$$

The empty set $\varnothing$ has a characteristic function that is identically equal to zero, $\varnothing(x) = 0$ for all x in **X**. The universe **X** itself comes with the characteristic function that is identically equal to one, that is, $X(x) = 1$ for all x in **X**. Also, a singleton A = {a}, a set comprising only a single element, has a characteristic function such that $A(x) = 1$ if x = a and $A(x) = 0$ otherwise.

Characteristic functions $A: X \rightarrow \{0, 1\}$ induce a constraint with well-defined binary boundaries imposed on the elements of the universe **X** that can be assigned to a set A. By looking at the characteristic function, we see that all elements belonging to the set are nondistinguishable—they come with the same value of the characteristic function so by knowing that $A(x_1) = 1$ and $A(x_2) = 1$, we cannot tell these elements apart. The operations of union, intersection, and complement are easily expressed in terms of the characteristic functions. The characteristic function of the union comes as the maximum of the characteristic functions of the sets involved in the operation. The complement of A denoted by $\bar{A}$, comes with a characteristic function equal to $1 - A(x)$.

Two-valued logic and sets are isomorphic. Logic statements (propositions) assume truth values of false (0) and true (1). Logic operations of *and* and *or* are isomorphic with the operations of union and intersection defined for sets. The negation flips the original truth value of the proposition.

Along with the introduction of sets is the concept of a relation. Relations represent and quantify associations between objects. They provide a fundamental vehicle to describe interactions and dependencies between variables, components, modules, nodes of the social network, and so forth. Relations capture associations between objects. Consider two spaces **X** and **Y**. Now form a Cartesian product of **X** and **Y**, that is, **X**×**Y**. The Cartesian product of **X** and **Y**, denoted **X**×**Y**, is the set of all pairs (x,y) such that x∈**X** and y∈**Y**. We define a relation R as the set of pairs of elements (x_i, y_j), $R = \{(x_i, y_j) \mid x_i \in X,$ and

$y_j \in$ **Y**}. In terms of the characteristic function, we express this as follows: $R(x_i, y_j) = 1$ if element x_i and y_i, are associated (related), and $R(x_i, y_j) = 0$ otherwise. For instance, in decision making, situations (x_i) and actions (y_j) are related: to each situation (state of nature) we assign a collection of pertinent actions, which are of interest.

More formally, a relation R defined over the Cartesian product of **X** and **Y**, is a collection of selected pairs (x,y) where $x \in$ **X** and $y \in$ **Y**. Equivalently, we write this down as

$$R: \mathbf{X} \times \mathbf{Y} \to \{0,1\} \tag{2.3}$$

The characteristic function of R is such that if $R(x,y) = 1$, then we say that the two elements x and y are related. If $R(x,y) = 0$, we say that these two elements (x and y) are unrelated. For example, suppose that **X** = **Y** = {1, 4, 8, 9}. The relation "equal to" formed over **X** × **X** is the set of pairs R = {$(x,y) \in$ **X** × **X** | x = y} = {(2,2), (4,4), (6,6), (8,8), (9,9)}. Its characteristic function is equal to

$$R(x, y) = \begin{cases} 1 & \text{if } x = y \\ 0 & \text{otherwise} \end{cases} \tag{2.4}$$

If a certain mapping "f" is a function, there is no guarantee that the mapping $f^{-1}: \mathbf{Y} \to \mathbf{X}$ is also a function, except in some cases when f^{-1} exists. Mappings are inherently linked with a concept of *direction*: we map from a certain space **X** to another one **Y**. In contrast, relations are direction-*free* constructs as there is no specific direction between the variables in X and Y identified. Being more descriptive, relations can be accessed from any direction. This makes a significant conceptual and computational difference in comparison with functions.

When a space under discussion involves "n" universes as its coordinates, an n-ary relation is any subset of the Cartesian product of these universes,

$$R: \mathbf{X}_1 \times \mathbf{X}_2 \times \dots \times \mathbf{X}_n \to \{0,1\} \tag{2.5}$$

2.2 Interval Analysis

Interval analysis has emerged with the inception of digital computers and was mostly motivated by the models of computations therein, which are carried out for intervals implied by the finite number of bits used to represent any number on a (digital) computer. This interval nature of the arguments (variables) implies that the results are also intervals. This raises awareness about the interval character of the results. Interval analysis is instrumental in the analysis of propagation of granularity residing within the original arguments (intervals).

Here, our intent is to elaborate on the fundamentals of interval calculus. It will become apparent that they will be helpful in the development of the algorithmic fabric of other formalisms of information granules.

We briefly recall the fundamental notions of numeric intervals. Two intervals $A = [a, b]$ and $B = [c, d]$ are equal if their bounds are equal, $a = c$ and $b = d$. A degenerate interval $[a,a]$ is a single number. There are two categories of operations on intervals (Alefeld and Herzberger, 1983; Moore, 1966; Moore, Kearfott, and Cloud, 2009), namely, set-theoretic and algebraic operations.

Set-theoretic operations. Assuming that the intervals are not disjoint (have some common points), they are defined as follows:

$$\text{Intersection } \{z \mid z \in A \text{ and } z \in B\} = [\max(a, c) \; \min(b, d)]$$

$$\text{Union } \{z \mid z \in A \text{ or } z \in B\} = [\min(a, c), \max(b, d)] \qquad (2.6)$$

For the illustration of the operations, refer to Figure 2.2.

Algebraic operations on intervals. The generic algebraic operations on intervals are quite intuitive. As before, let us consider the two intervals A and B. The results of addition, subtraction, multiplication, and division are expressed as follows:

$$A + B = [a+c, b+d]$$

$$A-B = [a-d, c-b] = A + [-1 \; -1]^* \, B$$

$$A^*B = [\min (ac, ad, bc, bd), \max (ac, ad, bc, bd)]$$

$$A/B = [a,b]^* \, [1/d, 1/c] \text{ (it is assumed that 0 is}$$

$$\text{not included in the interval } [c, d]) \qquad (2.7)$$

All these formulas result from the fact that the above functions are continuous on a compact set; as a result they take on the largest and the smallest

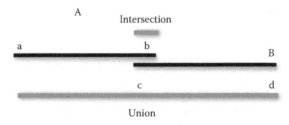

FIGURE 2.2
Examples of set-theoretic operations on numeric intervals.

value as well as the values in-between. The intervals of the obtained values are closed—in all these formulas, one computes the largest and the smallest values.

In addition to the algebraic operations, for a continuous unary operation $f(x)$ on **R** the mapping of the interval A produces an interval $f(A)$,

$$f(A) = [\min f(x), \max f(x)] \tag{2.8}$$

where the minimum (maximum) are taken for all x belonging to A. Examples of such mappings are x^k, $\exp(x)$, $\sin(x)$, and so forth.

Distance between intervals. The distance between two intervals A and B is expressed as

$$d(A,B) = \max (|a\text{-}c|, |b\text{-}d|) \tag{2.9}$$

One can easily show that the properties of distances are satisfied: $d(A, B) = d(B, A)$ (symmetry), $d(A,B)$ is nonnegative with $d(A, B) = 0$ (nonnegativity) if and only if $A = B$, $d(A, B) \le d(A,C) + d(B,C)$ (triangle inequality). For real numbers, the distance reduces to the Hamming one.

The metric for real intervals is the Hausdorff metric. Given that $f(x,y) = |x\text{-}y|$ and A and B are compact, nonempty sets of real numbers, the Hausdorff distance is defined in the form (Moore, 1966)

$$D(A, B) = \max[\sup_{v \in A} \inf_{u \in B} d(u,v), \sup_{u \in B} \inf_{v \in A} d(u,v)] \tag{2.10}$$

For the intervals A and B, the distance presented above is a special case of the Hausdorff distance.

An interval A defined in **X** mapped by a function (f) or a relation (R) gives rise to another interval B as shown in Figure 2.3.

We can express the mapping by the relation as a max-min composition of A and R, $B = A \circ R$, where the characteristic function of B reads as

$$B(y) = \max x_{\in X} [\min (A(x), R(x, y)] \tag{2.11}$$

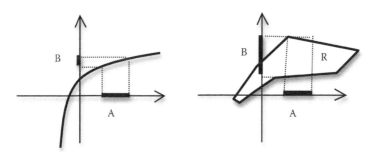

FIGURE 2.3
Mapping of interval via (a) function and (b) relation.

The above expression has a straightforward interpretation underlining how the mapping is realized. Without affecting the result, we start with the cylindric extension of A to the Cartesian product of X and Y c(A)(x,y) = A(x). Now c(A) and R are of the same nature—two relations defined in the Cartesian product of the same spaces. These two relations are intersected and then the result is projected on the space **Y**. The projection operation is realized by the maximum operation hence this operation is used in the above composition operation.

2.3 Fuzzy Sets: A Departure from the Principle of Dichotomy

Conceptually and algorithmically, fuzzy sets constitute one of the most fundamental and influential notions in science and engineering. The notion of a fuzzy set is highly intuitive and transparent since it captures what really becomes in essence a way in which the real world is being perceived and described in our everyday activities. We are faced with objects whose belongingness to a given category (concept) is always a matter of degree. There are numerous examples, in which we encounter elements whose allocation to the concept we want to define can be satisfied to some degree. One may eventually claim that continuity of transition from full belongingness and full exclusion is the major and ultimate feature of the physical world and natural systems. For instance, we may qualify an indoor environment as *comfortable* when its temperature is kept *around* 20°C. If we observe a value of 19.5°C it is very likely we still feel quite *comfortable*. The same holds if we encounter 20.5°C. Humans usually do not discriminate between changes in temperature within the range of one degree Celsius. A value of 20°C would be fully compatible with the concept of *comfortable* temperature yet 0°C or 30°C would not. In these two cases as well for temperatures close to these two values, we would describe them as being *cold* and *warm*, respectively. We could question whether the temperature of 25°C is viewed as *warm* or *comfortable* or, similarly, if 15°C is *comfortable* or *cold*. Intuitively, we know that 25°C is somehow between *comfortable* and *warm* while 15°C is between *comfortable* and *cold*. The value 25°C is partially compatible with the term *comfortable* and *warm*, and somewhat compatible or, depending on the observer's perception, incompatible with the term of *cold* temperature. Similarly, we may say that 15°C is partially compatible with the *comfortable* and *cold* temperature, and slightly compatible or incompatible with the warm temperature. In spite of this highly intuitive and apparent categorization of environmental temperatures into the three classes, namely, *cold*, *comfortable*, and *warm*, we note that the transition between the classes is not instantaneous and sharp (binary). Simply, when moving across the range of temperatures, these values become gradually perceived as *cold*, *comfortable*, or *warm*. A similar phenomenon happens when we are dealing with the concept of the height of people. An individual of a height of 1 meter is *short* whereas a person of 1.90 m is perceived to be *tall*. Again the question is, what is the range of height values that could qualify a person to be *tall*? Does a height of 1.85 m discriminate between *tall* and *short* individuals? Or maybe 1.86 m would be the right choice? Asking these questions, we sense that they do not make too much sense. We realize that the nature of these concepts is such that we cannot use a single number—a transition between the notion of tall and short is no abrupt in any way. Hence, we cannot assign a single number that does a good job. These simple and apparent instances send a clear message: the concept of dichotomy does not apply when defining even simple concepts. The reader can easily supply a huge number of examples as they commonly appear in natural language.

Fuzzy sets and the corresponding membership functions form a viable and mathematically sound framework to formalize these concepts. When talking about the heights of Europeans we may refer to real numbers within the interval [0, 3] to represent a universe of heights that range in-between 0 and 3 meters. This universe of discourse (space) is suitable for describing the concept of *tall* people.

The fundamental idea of a fuzzy set is to relax this requirement by admitting intermediate values of class membership. Therefore, we may assign intermediate values between 0 and 1 to quantify our perception of how compatible these values are with the class (concept) with 0 meaning incompatibility (complete exclusion) and 1 compatibility (complete membership). Membership values thus express the degrees to which each element of the universe is compatible with the properties distinctive to the class. Intermediate membership values underline that no "natural" threshold exists and that elements of a universe can be members of a class and at the same time belong to other classes with different degrees. Allowing for gradual, hence less strict nonbinary membership degrees is the crux of fuzzy sets.

Formally, a fuzzy set A is described by a membership function mapping the elements of a universe $\mathbf{X}$ to the unit interval [0,1] (Zadeh, 1965, 1975, 1978):

$$A: \mathbf{X} \rightarrow [0,1] \tag{2.12}$$

The membership functions are therefore synonymous with fuzzy sets. In a nutshell, membership functions generalize characteristic functions in the same way as fuzzy sets generalize sets.

The choice of the unit interval for the values of membership degrees is usually a matter of convenience. The specification of the very precise membership values (up to several decimal digits), that is, $A(4) = 0.9865$, is not crucial or even counterproductive. We should stress here that in describing membership grades we are predominantly after a reflection of an order of the elements in A in terms of their belongingness to the fuzzy set (Dubois and Prade, 1977, 1979, 1998).

Being more descriptive, we may view fuzzy sets as elastic constraints imposed on the elements of a universe. As emphasized before, fuzzy sets deal primarily with the concept of elasticity, graduality, or an absence of sharply defined boundaries. In contrast, when dealing with sets we are concerned with rigid boundaries, lack of graded belongingness, and sharp, binary boundaries. Gradual membership means that no natural boundary exists and that some elements of the universe of discourse can, contrary to sets, coexist (belong) to different fuzzy sets with different degrees of membership.

2.3.1 Membership Functions and Classes of Fuzzy Sets

Formally speaking, any function $A: \mathbf{X} \rightarrow [0, 1]$ could be qualified to serve as a membership function describing the corresponding fuzzy set. In practice, the form of the membership functions should be reflective of the problem at hand for which we construct fuzzy sets. They should reflect our perception (semantics) of the concept to be represented and further used in problem solving, the level of detail we intend to capture, and the context, in which the fuzzy sets are going to be used. It is also essential to assess the type of fuzzy set from the standpoint of its suitability when handling the ensuing optimization procedures. It also needs to accommodate some additional requirements arising as a result of further needs optimization procedures such as differentiability of membership functions. Keeping these criteria in mind, we elaborate on the most commonly used categories of membership functions. All of them are defined in the universe of real numbers, that is $\mathbf{X} = \mathbf{R}$.

Triangular membership functions. The fuzzy sets are expressed by their piecewise linear segments described in the form

$$A(x,a,m,b) = \begin{cases} 0 & \text{if } x \leq a \\ \dfrac{x-a}{m-a} & \text{if } x \in [a,m] \\ \dfrac{b-x}{b-m} & \text{if } x \in [m,b] \\ 0 & \text{if } x \geq b \end{cases} \tag{2.13}$$

Using more concise notation, the above expression can be written down in the form $A(x,a,m,b) = max\{min[(x-a)/(m-a),(b-x)/(b-m)],0\}$. The meaning of the parameters is straightforward: "m" denotes a modal (typical) value of the fuzzy set while "a" and "b" are the lower and upper bounds, respectively. They could be sought as the extreme elements of the universe of discourse that delineate the elements belonging to A with non-zero membership degrees. Triangular fuzzy sets (membership functions) are the simplest possible models of membership functions. They are fully defined by only three parameters. As mentioned, the semantics is evident as the fuzzy sets are expressed on a basis of knowledge of the spreads of the concepts and their typical values. The linear change in the membership grades is the simplest possible model of membership one could think of.

Trapezoidal membership functions. They are piecewise linear functions characterized by four parameters, "a," "m," "n," and "b" each of which defines one of the four linear parts of the membership function. They assume the following form:

$$A(x) = \begin{cases} 0 & \text{if } x < a \\ \dfrac{x-a}{m-a} & \text{if } x \in [a,m] \\ 1 & \text{if } x \in [m,n] \\ \dfrac{b-x}{b-n} & \text{if } x \in [n,b] \\ 0 & \text{if } x > b \end{cases} \tag{2.14}$$

Using an equivalent notation, we can rewrite A as follows: $A(x,a,m,n,b) = max\{min[(x{-}a)/(m{-}a), 1, (b{-}x)/(b{-}n)],0\}$. Note that elements in [m, n] are nondistinguishable as this region is described by a characteristic function.

S-membership functions. These functions are of the form:

$$A(x) = \begin{cases} 0 & \text{if } x \leq a \\ 2\left(\dfrac{x-a}{b-a}\right)^2 & \text{if } x \in [a,m] \\ 1-2\left(\dfrac{x-b}{b-a}\right)^2 & \text{if } x \in [m,b] \\ 1 & \text{if } x > b \end{cases} \tag{2.15}$$

The point m = (a + b)/2 is referred to as the crossover point.

Gaussian membership functions. These membership functions are described by the following relationship:

$$A(x,m,) = \exp\left(-\dfrac{(x-m)^2}{\sigma^2}\right) \tag{2.16}$$

Gaussian membership functions are described by two important parameters. The modal value (m) represents the typical element of A while s denotes a spread of A. Higher values of s correspond to larger spreads of the fuzzy sets.

2.3.2 Selected Descriptors of Fuzzy Sets

Given the enormous diversity of potentially useful (semantically sound) membership functions, there are certain common characteristics (descriptors) that are conceptually and operationally qualified to capture the essence of the granular constructs represented in terms of fuzzy sets. In what follows, we provide a list of the descriptors commonly encountered in practice.

Normality: We say that the fuzzy set A is *normal* if its membership function attains 1, that is,

$$\sup_{x \in X} A(x) = 1 \qquad (2.17)$$

If this property does not hold, we call the fuzzy set *subnormal*. The supremum operation (sup) standing in the above expression is also referred to as a height of the fuzzy set A, $hgt(A) = \sup_{x \in X} A(x) = 1$.

The normality of A has a simple interpretation: by determining the height of the fuzzy set, we identify an element with the highest membership degree. The value of the height being equal to one states that there is at least one element in X whose typicality with respect to A is the highest one and which could be sought as fully compatible with the semantic category presented by A.

The normalization operation, Norm(A) is a transformation mechanism that is used to convert a subnormal, nonempty fuzzy set A into its normal counterpart. This is done by dividing the original membership function by the height of this fuzzy set, that is,

$$Norm(A) = \frac{A(x)}{hgt(A)} \qquad (2.18)$$

Support: Support of a fuzzy set A, denoted by Supp (A), is a set of all elements of X with nonzero membership degrees in A,

$$Supp(A) = \{x \in X \mid A(x) > 0\} \qquad (2.19)$$

Core: The core of a fuzzy set A, Core(A), is a set of all elements of the universe that are typical for A, i.e., they come with membership grades equal to 1,

$$Core(A) = \{x \in X \mid A(x) = 1\} \qquad (2.20)$$

While core and support are somewhat extreme (in the sense that they identify the elements of A that exhibit the strongest and the weakest linkages with A), we may be also interested in characterizing sets of elements that come with some intermediate membership degrees. A notion of a so-called α-cut offers here an interesting insight into the nature of fuzzy sets.

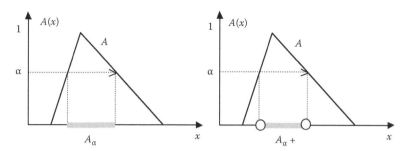

FIGURE 2.4
Examples of α-cut and strong α-cut.

α-cut: The α-cut of a fuzzy set A, denoted by A_α, is a set consisting of the elements of the universe whose membership values are equal to or exceed a certain threshold level α where $\alpha \in [0,1]$ (Zadeh, 1975; Nguyen and Walker, 1999; Pedrycz, Dong, and Hirota, 2009). Formally speaking, we have $A_\alpha = \{x \in X | A(x) \geq \alpha\}$. A strong α-cut differs from the α-cut in the sense that it identifies all elements in X for which we have the following equality $A = \{x \in X | A(x) > \alpha\}$. An illustration of the concept of the α-cut and strong α-cut is presented in Figure 2.4. Both support and core are limit cases of α-cuts and strong α-cuts. For $\alpha = 0$ and the strong α-cut, we arrive at the concept of the support of A. The threshold $\alpha = 1$ means that the corresponding α-cut is the core of A.

We can characterize fuzzy sets by counting their elements and bringing a single numeric quantity as a meaningful descriptor of this count. While in the case of sets this sounds convincing, here we have to take into account different membership grades. In the simplest form this counting comes under the name of cardinality.

Cardinality: Given a fuzzy set A defined in a finite or countable universe X, its cardinality, denoted by card(A), is expressed as the following sum,

$$\text{card}(A) = \sum_{x \in X} A(x) \tag{2.21}$$

or, alternatively, as the following integral,

$$\text{card}(A) = \int_X A(x)dx \tag{2.22}$$

(We assume that the integral shown above does make sense.) The cardinality produces a count of the number of elements in the given fuzzy set. As there are different degrees of membership, the use of the sum here makes sense as we keep adding contributions coming from the individual elements of this fuzzy set. Note that in the case of sets, we count the number of elements belonging to the corresponding sets. We also use the alternative notation of Card(A) = $|A|$, and refer to it as a sigma count (σ-count).

The cardinality of fuzzy sets is explicitly associated with the concept of granularity of information granules realized in this manner. More descriptively, the more elements of A we encounter, the higher the level of abstraction supported by A and the lower the granularity of the construct. Higher values of cardinality come with the higher level of abstraction (generalization) and the lower values of granularity (specificity). The concept of cardinality is linked to the notion of specificity of an information granule.

Specificity: Quite often, we face the issue of quantifying to what extent a single element of the universe could be regarded as representative of a fuzzy set defined there. If this fuzzy set is a singleton,

$$A(x) = \begin{cases} 1 \text{ if } x = x_0 \\ 0 \text{ if } x \neq x_0 \end{cases} \tag{2.23}$$

then there is no hesitation in selecting x_0 as the sole representative of A. In this case, we state that A is very *specific* and its choice comes with no hesitation. On the other extreme, if A covers the entire universe **X** and embraces all elements with the membership grade being equal to 1, the choice of the only one representative of A comes with a great deal of hesitation, which is triggered by a lack of specificity being faced in this problem. With this regard, refer also to the principle of the least commitment. These two extreme situations are portrayed in Figure 2.5. Intuitively, we sense that the specificity is a concept that relates quite visibly with the cardinality of a set (Yager, 1983). The higher the cardinality of the set is (i.e., the more evident its abstraction), the lower its specificity. Having said that, we are interested in developing a measure, which could be able to capture this effect of hesitation.

So far we discussed properties of a single fuzzy set. The operations to be studied look into the characterizations of relationships between two fuzzy sets.

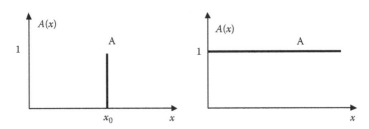

FIGURE 2.5
Examples of two extreme cases of sets exhibiting distinct levels of specificity.

Equality: We say that two fuzzy sets A and B defined in the same universe **X** are equal if and only if their membership functions are identical, meaning that

$$A(x) = B(x) \; \forall x \in \mathbf{X} \tag{2.24}$$

Inclusion: Fuzzy set *A* is a subset of *B* (*A* is included in *B*), denoted by $A \subseteq B$, if and only if every element of *A* also is an element of *B*. This property expressed in terms of membership degrees means that the following inequality is satisfied:

$$A(x) \le B(x) \; \forall x \in \mathbf{X} \tag{2.25}$$

2.3.3 Fuzzy Sets as a Family of α-Cuts

Fuzzy sets offer an important conceptual and operational feature of information granules by endowing their formal models by gradual degrees of membership. We are interested in exploring relationships between fuzzy sets and sets. While sets come with the binary (yes-no) model of membership, it could be worth investigating whether there are indeed some special cases of fuzzy sets and if so, in which sense a set could be treated as a suitable approximation of some given fuzzy set. This could shed light on some related processing aspects. To gain a detailed insight into this matter, we recall here the concept of an α-cut and a family of α-cuts and show that they relate to fuzzy sets in an intuitive and transparent way. Let us revisit the semantics of α-cuts: an α-cut of *A* embraces all elements of the fuzzy set whose degrees of belongingness (membership) to this fuzzy set are at least equal to α. In this sense, by selecting a sufficiently high value of α, we identify (tag) elements of A that belongs to it to a significant extent and thus could be sought as those substantially representative of the concept conveyed by A. Those elements of **X** exhibiting lower values of the membership grades are suppressed so this allows us to selectively focus on the elements with the highest degrees of membership while dropping the others.

For α-cuts A_α, the following properties hold:

(a) $A_0 = \mathbf{X}$

(b) If $\alpha \le \beta$ then $A_\alpha \supseteq A_\beta$ (2.26)

The first property tells us that if we allow for the zero value of α, then all elements of **X** are included in this α-cut (0-cut, to be more specific). The second property underlines the monotonic character of the construct: higher values of the threshold imply that less elements are accepted in the resulting α-cuts. In other words, we may say that the level sets (α-cuts) A_α form a nested family of sets indexed by some parameter (α). If we consider the limit value of α, that is $\alpha = 1$, the corresponding α-cut is nonempty if and only if A is a normal fuzzy set.

It is also worthwhile to remember that α-cuts, in contrast to fuzzy sets, are sets. We showed how for some given fuzzy set, its α-cut could be formed. An interesting question arises as to the construction that could be realized when moving in the opposite direction. Could we "reconstruct" a fuzzy set on the basis of an infinite family of sets? The answer to this problem is offered in what is known as the *representation theorem for fuzzy sets* (Klir and Yuan, 1995).

Theorem

Let $\{A_\alpha\}$ $\alpha \in [0,1]$ be a family of sets defined in **X** such that they satisfy the following properties:

(a) $A_0 = \mathbf{X}$
(b) If $\alpha \le \beta$ then $A_\alpha \supseteq A_\beta$
(c) For the sequence of threshold values $\alpha_1 \le \alpha_2 \le \alpha_n$ such that $\lim_{n \to \infty} \alpha_n = \alpha$, we have $A_\alpha = \bigcap_{n=1}^{\infty} A_{\alpha_n}$ ∎

Then there exists a unique fuzzy set B defined in **X** such that $B_\alpha = A_\alpha$ for each $\alpha \in [0,1]$.

In other words, the representation theorem states that any fuzzy set A can be uniquely represented by an infinite family of its α-cuts. The following reconstruction expression shows how the corresponding α-cuts contribute to the formation of the corresponding fuzzy set

$$A = \bigcup_{\alpha>0} \alpha A_\alpha \qquad (2.27)$$

that is,

$$A(x) = \sup_{\alpha \in (0,1]} [\alpha A_a(x)] \qquad (2.28)$$

where A_α denotes the corresponding α-cut.

The essence of this construct is that any fuzzy set can be uniquely represented by the corresponding family of nested sets (namely, ordered by the inclusion relation). The illustration of the concept of the α-cut and a way in which the representation of the corresponding fuzzy set becomes realized becomes visualized in Figure 2.6.

More descriptively, we may say that fuzzy sets can be reconstructed with the aid of a family of sets. Apparently, we need a family of sets (intervals, in particular) to capture the essence of a single fuzzy set. The reconstruction

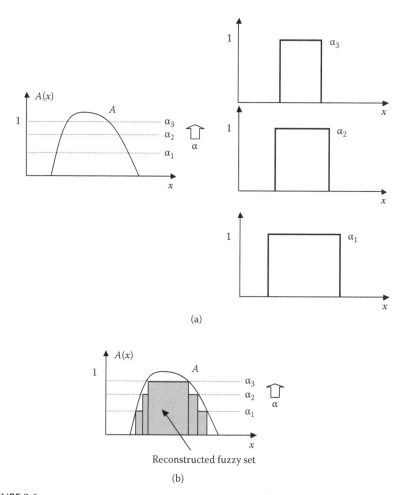

(a)

(b)

FIGURE 2.6
Fuzzy set A. (a) Examples of some of its α-cuts. (b) A representation of A through the corresponding family of sets (α-cuts).

scheme illustrated in Figure 2.6 is self-explanatory in this regard. In more descriptive terms, we may look at the expression offered by Equations (2.27) and (2.28) as a way of decomposing A into a series of layers (indexed sets) being calibrated by the values of the associated levels of α.

For the finite universe of discourse, dim $(\mathbf{X}) = n$, we encounter a finite number of membership grades and subsequently a finite number of α-cuts. This finite family of α-cuts is then sufficient to fully *represent* or reconstruct the original fuzzy set. From the practical perspective, it is helpful in the development and implementation of algorithms with fuzzy sets in case there are well-established algorithms for intervals. The algorithm is realized for several α-cuts and then the partial results are combined to produce a solution in the form of a fuzzy set.

2.3.4 Triangular Norms and Triangular Conorms as Models of Operations on Fuzzy Sets

Logic operations on fuzzy sets concern manipulation of their membership functions. Therefore they are domain dependent and different contexts may require their different realizations. For instance, since operations provide ways to combine information, they can be performed differently in image processing, control, and diagnostic systems, for example. When contemplating the realization of operations of intersection and union of fuzzy sets, we should require a satisfaction of the following collection of intuitively appealing properties: (a) commutativity, (b) associativity, (c) monotonicity, and (d) identity.

The last requirement of identity takes on a different form depending on the operation. In the case of intersection, we anticipate that an intersection of any fuzzy set with the universe of discourse **X** should return this fuzzy set. For the union operations, the identity implies that the union of any fuzzy set and an empty fuzzy set returns the fuzzy set.

Thus, any binary operator $[0,1] \times [0,1] \to [0,1]$, which satisfies the collection of the requirements outlined above, can be regarded as a potential candidate to realize the intersection or union of fuzzy sets. Note also that identity acts as boundary conditions meaning that when confined to sets, the above stated operations return the same results as encountered in set theory. In general, idempotency is not required, however the realizations of union and intersection could be idempotent as this happens for the operations of minimum and maximum where $\min(a, a) = a$ and $\max(a, a) = a$.

In the theory of fuzzy sets, triangular norms (Schweizer and Sklar, 1983; Klement, Mesiar, and Pap, 2000) offer a general class of operators of intersection and union. Originally, they were introduced in the context of probabilistic metric spaces. T-norms arise a family of operators modeling intersection of fuzzy sets. Given a t-norm, a dual operator called a *t-conorm* (or *s-norm*) can be derived using the relationship $x \, s \, y = 1 - (1 - x) \, t \, (1 - y)$, $\forall \, x,y \in [0,1]$, which is the De Morgan law. Triangular conorms provide generic models for the union of fuzzy sets. T-conorms can also be described by an independent axiomatic system (Valverde and Ovchinnikov, 2008).

A triangular norm, t-norm for brief, is a binary operation t: $[0,1] \times [0,1] \to [0,1]$ that satisfies the following properties:

Commutativity:	a t b = b t a
Associativity:	a t (b t c) = (a t b) t c
Monotonicity:	if b ≤ c then a t b ≤ a t c
Boundary conditions:	a t 1 = a
	a t 0 = 0

where a, b, c $\in$ [0,1].

Let us elaborate on the meaning of these requirements vis-à-vis the use of t-norms as models of operators of union and intersection of fuzzy sets. There is a one-to-one correspondence between the general requirements outlined above and the properties of t-norms. The first three reflect the general character of set operations. Boundary conditions stress the fact that all t-norms attain the same values at boundaries of the unit square $[0,1] \times [0,1]$. Thus, for sets, any t-norm produces the same result that coincides with the one we could have expected in set theory when dealing with intersection of sets, that is, $A \cap X = A$, $A \cap \varnothing = \varnothing$. Some commonly encountered examples of t-norms include the following operations:

$$
\begin{aligned}
\text{Minimum:} \quad & a \, t_m \, b = \min(a, b) = a \wedge b \\
\text{Product:} \quad & a \, t_p \, b = ab \\
\text{Lukasiewicz:} \quad & a \, t_l \, b = \max(a + b - 1, 0)
\end{aligned}
$$

$$
\text{Drastic product:} \quad a \, t_d \, b = \begin{cases} a & \text{if } b = 1 \\ b & \text{if } a = 1 \\ 0 & \text{otherwise} \end{cases}
$$

In general, t-norms cannot be linearly ordered. One can demonstrate though that the min (t_m) t-norm is the largest t-norm, while the drastic product is the smallest one. They form the lower and upper bounds of the t-norms in the following sense:

$$ a \, t_d \, b \leq a \, t \, b \leq a \, t_m \, b = \min(a, b) \tag{2.29} $$

Triangular conorms are functions s: $[0,1] \times [0,1] \rightarrow [0,1]$ that serve as generic realizations of the union operator on fuzzy sets. Similarly as triangular norms, conorms provide the highly desirable modeling flexibility needed to construct fuzzy models. Triangular conorms can be viewed as dual operators to the t-norms and as such, are explicitly defined with the use of De Morgan laws. We may characterize them in a fully independent fashion by offering the following definition.

A triangular conorm (s-norm) is a binary operation s: $[0,1] \times [0,1] \rightarrow [0,1]$ that satisfies the following requirements

$$
\begin{aligned}
\text{Commutativity:} \quad & a \, s \, b = b \, s \, a \\
\text{Associativity:} \quad & a \, s \, (b \, s \, c) = (a \, s \, b) \, s \, c \\
\text{Monotonicity:} \quad & \text{if } b \leq c \text{ then } a \, s \, b \leq a \, s \, c \\
\text{Boundary conditions:} \quad & a \, s \, 0 = a \\
& a \, s \, 1 = 1 \\
& a, b, c \in [0,1].
\end{aligned}
$$

One can show that s: $[0,1] \times [0,1] \rightarrow [0,1]$ is a t-conorm if and only if (iff) there exists a t-norm (dual t-norm) such that for $\forall$ a,b $\in$ [0,1] we have

$$a \, s \, b = 1 - (1 - a) \, t \, (1 - b) \tag{2.30}$$

For the corresponding dual t-norm, we have

$$a \, t \, b = 1 - (1 - a) \, s \, (1 - b) \tag{2.31}$$

The duality expressed by Equations (2.30) and (2.31) can lead to an alternative definition of t-conorms. This duality allows us to deduce the properties of t-conorms on the basis of the analogous properties of t-norms. Notice that after rewriting Equations (2.30) and (2.31), we obtain

$$(1 - a) \, t \, (1 - b) = 1 - a \, s \, b \tag{2.32}$$

$$(1 - a) \, s \, (1 - b) = 1 - a \, t \, b \tag{2.33}$$

These two relationships can be expressed symbolically as

$$\overline{A} \cap \overline{B} = \overline{A \cup B} \tag{2.34}$$

$$\overline{A} \cup \overline{B} = \overline{A \cap B} \tag{2.35}$$

which are nothing but the De Morgan laws well known in set theory. As shown, they are also satisfied for fuzzy sets.

The boundary conditions mean that all t-conorms behave similarly at the boundary of the unit square $[0,1] \times [0,1]$. Thus, for sets, any t-conorm returns the same result as that encountered in set theory.

A list of commonly used t-conorms includes the following examples:

Maximum: $a \, s_m \, b = \max (a, b) = a \vee b$

Probabilistic sum: $a \, s_p \, b = a + b - ab$

Lukasiewicz: $a \, s_l \, b = \min (a + b, 1)$

Drastic sum: $a \, s_d \, b = \begin{cases} a & \text{if } b=0 \\ b & \text{if } a=0 \\ 1 & \text{otherwise} \end{cases}$

2.4 Rough Sets

The description of information granules completed with the aid of some vocabulary is usually imprecise. Intuitively, such description may lead to some approximations, called *lower* and *upper bounds*. This is the essence of rough sets introduced by Pawlak (1982). Interesting generalizations, conceptual insights, and algorithmic investigations are offered in a series of fundamental papers by Pawlak and Skowron (2007a,b).

To explain the concept of rough sets and show what they have to offer in terms of representing information granules, we use an illustrative example. Consider a description of environmental conditions expressed in terms of temperature and pressure. For each of these factors, we fix several ranges of possible values where each of such ranges comes with some interpretation such as "values below," "values in-between," "values above," and so forth. By admitting such selected ranges in both variables, we construct a grid of concepts formed in the Cartesian product of the spaces of temperature and pressure (refer to Figure 2.7). In more descriptive terms, this grid forms a vocabulary of generic terms, which we would like to use to describe all new information granules.

Now let us consider that the environmental conditions monitored over some time have resulted in some values of temperature and

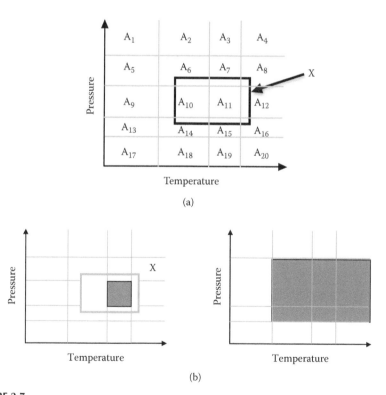

(a)

(b)

FIGURE 2.7
A collection of elements of the vocabulary and their use in the problem description. (a) Environmental conditions X result in some interval of possible values. In the sequel, this gives rise to the concept of a rough set with the roughness of the description being captured by the lower and upper bounds (approximations) as illustrated in (b).

pressure ranging in-between some lower and upper bound as illustrated in Figure 2.7. Denote this result by X. When describing it in the terms of the elements of the vocabulary, we end up with a collection of elements that are fully included in X. They form a lower bound of description of X when being completed in the presence of the given vocabulary. Likewise, we may identify elements of the vocabulary that have a nonempty overlap with X and, in this sense, constitute an upper bound of the description of the given environmental conditions. Along with the vocabulary, the description forms a certain rough set. The roughness of the resulting description is inherently associated with a finite size of the vocabulary. More formally, we describe an upper bound by enumerating elements of A_is that have a nonzero overlap with X, that is,

$$X_+ = \{A_i \mid A_i \cap X \neq \emptyset\} \qquad (2.36)$$

More specifically, as shown in Figure 2.7, we $X_+ = \{A_6, A_7, A_8, A_{10}, A_{11}, A_{12}, A_{14}, A_{15}, A_{16}\}$. The lower bound of X involves all A_i such that they are fully included within X, namely,

$$X_- = \{A_i \mid A_i \subset X\} \qquad (2.37)$$

Here $X_- = \{A_{11}\}$. As succinctly visualized in Figure 2.7, we are concerned with a description of a given concept X realized in the language of a certain collection (vocabulary) of rather generic and simple terms $A_1, A_2, ..., A_c$. The lower and upper boundaries (approximation) are reflective of the resulting imprecision caused by the conceptual incompatibilities between the concept itself and the existing vocabulary (see Figure 2.8).

The bounds as described above can also be expressed in the following form:

$$X_+(A_i) = \sup_x[\min(A_i(x), X(x))] = \sup_{x \in supp(A_i)} X(x) \qquad (2.38)$$

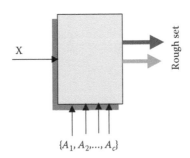

FIGURE 2.8
Rough set as a result of describing X in terms of some fixed vocabulary $\{A_1, A_2, ..., A_c\}$; the lower and upper bounds are results of the description.

Subsequently, the lower bound is expressed in the form

$$X_-(A_i) = \inf_x [\max(1 - A_i(x), X(x))] \tag{2.39}$$

It is interesting to note that the vocabulary used in the above construct could comprise information granules being expressed in terms of any other formalism, that is, fuzzy sets. Quite often we can encounter constructs like rough fuzzy sets and fuzzy rough sets in which both fuzzy sets and rough sets are put together. We discuss these constructs in detail in Chapter 3.

2.5 Shadowed Sets as a Three-Valued Logic Characterization of Fuzzy Sets

Fuzzy sets are associated with the collections of numeric membership grades. Shadowed sets (Pedrycz, 1998) are based upon fuzzy sets by forming a more general and highly synthetic view of the numeric concept of membership. Using shadowed sets, we quantify numeric membership values into three categories: complete belongingness, complete exclusion, and unknown (which could be also conveniently referred to as *don't know condition* or a *shadow*). A graphic illustration of a shadowed set along with the principles of sets and fuzzy sets is schematically shown in Figure 2.9. This helps us contrast these three fundamental constructs of information granules.

Shadowed sets reveal interesting linkages between fuzzy sets and sets, and show some relationships with rough sets.

In a nutshell, shadowed sets can be regarded as a general and far more concise representation of a fuzzy set that could be of particular interest when dealing with further computing (in which case we could come up with substantial reduction of the overall processing effort).

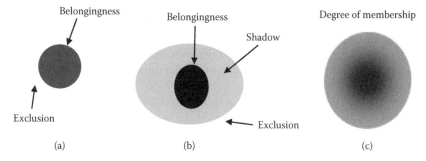

FIGURE 2.9
(a) A schematic view of sets. (b) Shadowed sets. (c) Fuzzy sets.

Fuzzy sets offer a wealth of detailed numeric information conveyed by their detailed numeric membership grades (membership functions). This very detailed conceptualization of information granules can clearly act as a two-edged sword. On one hand, we may enjoy a very detailed quantification of elements to a given concept (fuzzy set). On the other hand, those membership grades could be somewhat overwhelming and introduce some burden when it comes to a general interpretation. It is also worth noting that numeric processing of membership grades comes sometimes with quite substantial computing overhead. To alleviate these problems, we introduce a certain category of information granules called *shadowed sets*. Shadowed sets are information granules induced by fuzzy sets so that they capture the essence of fuzzy sets while at the same time reducing the numeric burden because of their limited three-valued characterization of shadowed sets. This nonnumeric character of shadowed sets is also of particular interest when dealing with their interpretation abilities. Given the characteristics of shadowed sets, we may view them as a certain symbolic representation of fuzzy sets.

2.5.1 Defining Shadowed Sets

Formally speaking, a shadowed set A defined in some space $\mathbf{X}$ is a set-valued mapping coming in the following form (Pedrycz, 1998):

$$A : \mathbf{X} \rightarrow \{0, [0,1], 1\} \qquad\qquad (2.40)$$

The codomain of A consists of three components that is 0, 1, and the unit interval [0,1]. They can be treated as degrees of membership of elements to A. These three quantification levels come with an apparent interpretation. All elements for which A(x) assume 1 are called a *core of the shadowed set*—they embrace all elements that are fully compatible with the concept conveyed by A. The elements of $\mathbf{X}$ for which A(x) attains zero are excluded from A. The elements of $\mathbf{X}$ for which we have assigned the unit interval are completely uncertain—we are not in a position to allocate any numeric membership grade. Therefore, we allow the usage of the unit interval, which reflects uncertainty meaning that any numeric value could be permitted here. In essence, such element could be excluded (we pick up the lowest possible value from the unit interval), exhibit partial membership (any number within the range from 0 and 1) or could be fully allocated to A. Given this extreme level of uncertainty (nothing is known and all values are allowed), we call these elements *shadows* and hence the name of the shadowed set. Alternatively, we can stress this quantification by using the notation: full membership $A^{+} = \{x \in \mathbf{X} | x \in A\}$, exclusion $A^{-} = \{x \in \mathbf{X} | x \notin A\}$, and ignorance about membership (shadow): $A^{?} = \{x \in \mathbf{X} | x?A\}$. An illustration of the underlying concept of a shadowed set is included in Figure 2.10.

One can view this mapping (shadowed set) as an example of a three-valued logic as encountered in the classic model introduced by Lukasiewicz.

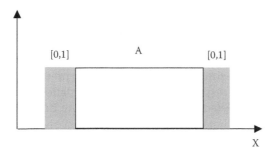

FIGURE 2.10
An example of a shadowed set A. Note shadows formed around the cores of the construct.

Having this in mind, we can think of shadowed sets as a *symbolic* representation of *numeric* fuzzy sets. Obviously, the elements of the codomain of A could be labeled using symbols (i.e., certain, shadow, excluded; or *a, b, c,* and the like) endowed with some well-defined semantics.

The operations on shadowed sets (Table 2.1) are isomorphic with those encountered in the three-valued logic.

These logic operations are conceptually convincing; we observe an effect of preservation of uncertainty. In the case of the *or* operation, we note that combining a single numeric value of exclusion (0) with the shadow, we arrive

TABLE 2.1

Logic Operations (*And, Or,* and Complement) on Shadowed Sets

A \ B	0	S	1
0	0	0	0
S	0	S	S
1	0	S	1

Intersection

A \ B	0	S	1
0	0	S	1
S	S	S	1
1	1	1	1

Union

A			
0	1		
S	S		
1	0		

Complement

Note: Here S stands for a shadow, S (= [0,1]).

at the shadow (as nothing specific could be stated about the result of this logic aggregation). A similar effect occurs for the *and* operator when applied to the shadow and the logic value of 1.

The simplicity of shadowed sets becomes their obvious advantage. Dealing with three logic values simplifies not only the interpretation but it is advantageous in all computing, especially when such calculations are contrasted with the calculations completed for fuzzy sets involving detailed numeric membership grades. Let us note that logic operators that are typically realized by means of some t-norms and t-conorms require computing of the numeric values of the membership grades. In contrast those realized on shadowed sets are based on comparison operations and therefore are far less demanding.

While shadowed sets could be sought as new and standalone constructs, our objective is to treat them as information granules induced by some fuzzy sets. The bottom line of our approach is straightforward—considering fuzzy sets (or fuzzy relations) as the point of departure and acknowledging the computing overhead associated with them, we regard shadowed sets as constructs that capture the essence of fuzzy sets while helping to reduce the overall computing effort and simplifying the ensuing interpretation. In the next section, we concentrate on the development of shadowed sets for given fuzzy sets.

2.5.2 The Development of Shadowed Sets

Accepting the point of view that shadowed sets are algorithmically implied (induced) by some fuzzy sets, we are interested in the transformation mechanisms translating fuzzy sets into the corresponding shadowed sets. The underlying concept is the one of uncertainty condensation or "localization." While in fuzzy sets we encounter intermediate membership grades located in-between 0 and 1 and distributed practically across the entire space, in shadowed sets we "localize" the uncertainty effect by building constrained and fairly compact shadows. By doing so we could remove (or better to say, redistribute) uncertainty from the rest of the universe of discourse by bringing the corresponding low and high membership grades to zero and one, and then compensating these changes by allowing for the emergence of uncertainty regions. This transformation could lead to a certain optimization process in which we complete a total balance of uncertainty.

To illustrate this optimization, let us start with a continuous, symmetric, unimodal, and normal membership function A. In this case, we can split the problem into two tasks by considering separately the increasing and decreasing portion of the membership function (Figure 2.11).

For the increasing portion of the membership function, we reduce low membership grades to zero, elevate high membership grades to one and compensate these changes (which in essence lead to an elimination of partial membership grades) by allowing for a region of the shadow where there are no specific membership values assigned but we admit the entire unit interval as feasible membership grades. Computationally, we form the

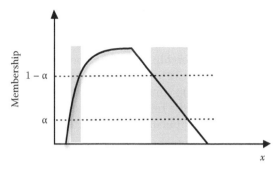

FIGURE 2.11
The concept of a shadowed set induced by some fuzzy set. Note the range of membership grades (located between α and 1- α) generating a shadow.

following balance of uncertainty preservation that could be symbolically expressed as

$$\text{Reduction of membership + Elevation of membership = Shadow} \quad (2.41)$$

Again referring to Figure 2.11, once given the membership grades below α and above 1– α, α ∈ (0, ½), we express the components of the above relationship in the form (we assume that all integrals do exist)

Reduction of membership (low membership grades are reduced to zero),

$$\int_{x:A(x)\leq\alpha} A(x)dx \quad (2.42)$$

Elevation of membership (high membership grades elevated to 1),

$$\int_{x:A(x)\geq 1-\alpha} (1-A(x))dx \quad (2.43)$$

$$\text{Shadow,} \quad \int_{x:\alpha<A(x)<1-\alpha} dx \quad (2.44)$$

The minimization of the absolute difference

$$V(\alpha) = \left| \int_{x:A(x)\leq\alpha} A(x)dx + \int_{x:A(x)\geq 1-\alpha} (1-A(x))dx + \int_{x:\alpha<A(x)<1-\alpha} dx \right| \quad (2.45)$$

completed with respect to α is given in the form of the following optimization problem:

$$\alpha_{opt} = \arg\min_{\alpha} V(\alpha) \quad (2.46)$$

where $\alpha \in (0, \frac{1}{2})$. For instance, when dealing with triangular membership function (and it appears that the result does not require the symmetry requirement), the optimal value of α is equal to $\sqrt{2} - 1 \approx 0.4142$ (Pedrycz, 1998). For the parabolic membership functions, the optimization leads to the value of α equal to 0.405.

For the Gaussian membership function, $A(x) = \exp(-x^2/\sigma^2)$, we get the optimal value of α resulting from the relationship (the calculations here concern the decreasing part of the membership function defined over $[0, \infty]$

$$V(\alpha) = | \int_0^{\sigma\sqrt{-\ln(1-\alpha)}} (1 - A(x))dx + \int_{\sigma\sqrt{-\ln(\alpha)}}^{\infty} A(x)dx - \int_{\sigma\sqrt{-\ln(1-\alpha)}}^{\sigma\sqrt{-\ln(\alpha)}} dx | \qquad (2.47)$$

Then the optimal value of α is equal to 0.3947 and it does not depend upon the spread σ.

2.6 Conclusions

Information granules are constructed with a well-defined semantics, which help formalize a notion of abstraction. Different formal approaches emphasize this fundamental facet in different ways. Sets stress the notion of dichotomy. Fuzzy sets depart from the dichotomy by emphasizing the idea of partial membership and in this way offer a possibility to deal with concepts where the binary view of the underlying concept is not suitable or overly restrictive. Rough sets bring another conceptual and algorithmic perspective by admitting an inability for a full description of concepts in the presence of a limited collection of information granules. Roughness comes as a manifestation of the limited descriptive capabilities of the vocabulary used, with which the description is realized. Shadowed sets can be sought as a concise, less numerically driven characterization of fuzzy sets (and induced by fuzzy sets themselves). One could also take another view of them as forming a bridge between fuzzy sets and rough sets. In essence, the shadows can be regarded as regions where uncertainty has been accumulated.

The choice of a suitable formal framework of information granules is problem oriented and has to be done according to the requirements of the problem, ways of acquisition of information granules (both on the basis of domain knowledge and numerical evidence), and related processing and optimization mechanisms available in light of the assumed formalism of information granularity.

References

Alefeld, G. and J. Herzberger. 1983. *Introduction to Interval Computations*. New York: Academic Press.

Dubois, D. and H. Prade. 1979. Outline of fuzzy set theory: An introduction. In: *Advances in Fuzzy Set Theory and Applications*, M.M. Gupta, R.K. Ragade, and R.R. Yager (eds.). Amsterdam: North-Holland. pp. 27–39.

Dubois, D. and H. Prade. 1977. The three semantics of fuzzy sets. *Fuzzy Sets and Systems*, 90, 141–150.

Dubois, D. and H. Prade. 1998. An introduction to fuzzy sets. *Clinica Chimica Acta*, 70, 3–29.

Klement, P., R. Mesiar, and E. Pap. 2000. *Triangular Norms*. Dordrecht: Kluwer Academic Publishers.

Klir, G. and B. Yuan. 1995. *Fuzzy Sets and Fuzzy Logic: Theory and Applications*. Upper Saddle River: Prentice-Hall.

Moore, R. 1966. *Interval Analysis*. Englewood Cliffs: Prentice Hall.

Moore, R., R.B. Kearfott, and M.J. Cloud. 2009. *Introduction to Interval Analysis*. Philadelphia: SIAM.

Nguyen, H. and E. Walker. 1999. *A First Course in Fuzzy Logic*. Boca Raton: Chapman Hall, CRC Press.

Pawlak, Z. 1982. Rough sets. *International Journal of Information and Computer Science*, 11, 15, 341–356.

Pawlak, Z. and A. Skowron. 2007a. Rudiments of rough sets. *Information Sciences*, 177, 1, 1, 3–27.

Pawlak, Z. and A. Skowron. 2007b. Rough sets and Boolean reasoning. *Information Sciences*, 177, 1, 41–73.

Pedrycz, W. 1998. Shadowed sets: Representing and processing fuzzy sets. *IEEE Trans. on Systems, Man, and Cybernetics, Part B*, 28, 103–109.

Pedrycz, A., F. Dong, and K. Hirota. 2009. Finite α cut-based approximation of fuzzy sets and its evolutionary optimization. *Fuzzy Sets and Systems*, 160, 3550–3564.

Schweizer, B. and A. Sklar. 1983. *Probabilistic Metric Spaces*. New York: North-Holland.

Valverde, L. and S. Ovchinnikov. 2008. Representations of T-similarity relations. *Fuzzy Sets and Systems*, 159, 2211–2220.

Yager, R. 1983. Entropy and specificity in a mathematical theory of evidence. *International Journal of General Systems*, 9, 249–260.

Zadeh, L.A. 1965. Fuzzy sets. *Information and Control*, 8, 338–353.

Zadeh, L.A. 1975. The concept of linguistic variables and its application to approximate reasoning I, II, III. *Information Sciences*, 8, 199–249, 301–357, 43–80.

Zadeh, L.A. 1978. Fuzzy sets as a basis for a theory of possibility. *Fuzzy Sets and Systems*, 1, 3–28.

3

Information Granules of Higher Type and Higher Order, and Hybrid Information Granules

This chapter provides an introduction to more advanced, hierarchically structured information granules such as those of higher type and higher order. In general, when talking about information granules of higher type, that is, type-2, we mean information granules whose elements are characterized by membership grades, which themselves are information granules (instead of being plain numeric values). For instance, in type-2 fuzzy sets, membership grades are quantified as fuzzy sets in [0,1] or intervals in the unit interval. Of course, one could envision a plethora of the constructs along this line; for instance, the membership grades could be rough sets or probability density functions (as this is the case in probabilistic sets). When talking about higher-order information granules, we mean constructs for which the universe of discourse comprises a family of information granules instead of single elements. Hybridization, on the other hand, is about bringing several formalisms of information granules and using them in an orthogonal setting. This situation is visible in fuzzy probabilities.

3.1 Fuzzy Sets of Higher Order

In Chapter 1, we noted that there is an apparent distinction between *explicit* and *implicit* manifestations of fuzzy sets. This observation triggers further conceptual investigations and leads to the concept of fuzzy sets of higher order. Let us recall that a fuzzy set is defined in a certain universe of discourse **X** so that for each element of it we come up with the corresponding membership degree, which is interpreted as the degree of compatibility of this specific element with the concept conveyed by the fuzzy set under discussion. The essence of a fuzzy set of second order is that it is defined over a collection of some other generic fuzzy sets. As an illustration, let us consider the concept of a *comfortable* temperature which we define over a finite collection of some generic fuzzy sets, that is, *around* 10°C, *warm*, *hot*, *cold*, *around* 20°C,... and so forth. We could easily come to a quick conclusion that the term *comfortable* sounds more *descriptive* and hence becomes semantically

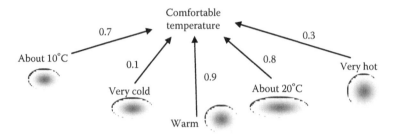

FIGURE 3.1

An example of a second-order fuzzy set of *comfortable* temperature defined over a collection of basic terms—generic fuzzy sets (graphically displayed as small clouds). Corresponding membership degrees are also shown.

more advanced being in rapport with real-world concepts in comparison to the generic terms which we used to describe it. An illustration of this second-order fuzzy set is illustrated in Figure 3.1. To make a clear distinction, fuzzy sets studied so far could be referred to as fuzzy sets of the first order.

Using the membership degrees as portrayed in Figure 3.1, we can write down the membership of *comfortable* temperature in the vector form as [0.7 0.1 0.9 0.8 0.3]. It is understood that the corresponding entries of this vector pertain to the generic fuzzy sets we started with when forming the fuzzy set. Figure 3.2 graphically emphasizes the difference between fuzzy sets (fuzzy

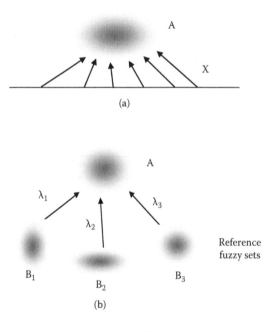

FIGURE 3.2

Contrasting fuzzy sets of order 1 (a) and order 2 (b). Note the role reference fuzzy sets played in the development of order-2 fuzzy sets.

sets of the first order) and fuzzy sets of the second order. For the order-2 fuzzy set, we can use the notation B = [λ_1, λ_2, λ_3] given that the reference fuzzy sets are A_1, A_2, and A_3, and the corresponding entries of B are the numeric membership grades specifying to which extent the reference fuzzy sets form the higher-order concept.

Fuzzy sets of order 2 could be also formed on a Cartesian product of some families of generic fuzzy sets. Consider, for instance, a concept of a *preferred* car. To everybody this term could mean something different, yet all of us agree that the concept itself is quite complex and definitely multifaceted. We easily include several aspects such as economy, reliability, depreciation, acceleration, and others. For each of these aspects we might have a finite family of fuzzy sets, for example, when talking about economy; we may use descriptors such as *about* 10 1/100 km (or expressed in mpg), *high* fuel consumption, *about* 30 mpg, and so forth. For the given families of generic fuzzy sets in the vocabulary of generic descriptors we combine them in a hierarchical manner as illustrated in Figure 3.3.

In a similar way, we can propose fuzzy sets of higher order such as third order or higher. They are formed in a recursive manner. While conceptually appealing and straightforward, their applicability could become an open issue. One may not venture to allocate more effort into their design unless there is a legitimate reason behind the further usage of fuzzy sets of higher order.

Nothing prevents us from building fuzzy sets of second order on a family of generic terms that are not only fuzzy sets. One might consider a family of information granules such as sets over which a certain fuzzy set is being formed.

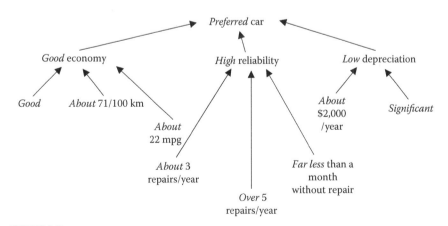

FIGURE 3.3
Fuzzy set of order 2 for *preferred* cars. Note the number of descriptors quantified in terms of fuzzy sets and contributing directly to their formation.

3.2 Rough Fuzzy Sets and Fuzzy Rough Sets

It is interesting to note that the vocabulary used in the above construct could comprise information granules expressed in terms of any other formalism such as fuzzy sets. Quite often we can encounter constructs like rough fuzzy sets and fuzzy rough sets in which both fuzzy sets and rough sets are organized together.

These constructs rely on the interaction between fuzzy sets and sets being used in the construct. Let us consider a finite collection of sets $\{A_i\}$ and use them to describe some fuzzy set X. In this scheme, we arrive at the concept of a certain fuzzy rough set (refer to Figure 3.4). The upper bound of this fuzzy rough set is computed as in the previous case yet given the membership function of X, the detailed calculations return membership degrees rather than 0-1 values. Given the binary character of A_is, the above expression for the upper bound comes in the form

$$X_+(A_i) = \sup_x[\min(A_i(x), X(x))] = \sup_{x \in \text{supp}(A_i)} X(x) \tag{3.1}$$

The lower bound of the resulting fuzzy rough set is taken in the form

$$X_-(A_i) = \inf_x[\max(1-A_i(x), X(x))] \tag{3.2}$$

Example

Let us consider a universe of discourse $\mathbf{X} = [-3, 3]$ and a collection of intervals regarded as a family of basic granular descriptors (see Figure 3.5).

The fuzzy set A with a triangular membership function distributed between -2 and 2 gives rise to some rough set with the lower and upper approximation of the form

$X_+ = [0.0\ 0.5\ 1.0\ 1.0\ 0.5\ 0.0]$ and $X_- = [0.0\ 0.0\ 0.5\ 0.5\ 0.0\ 0.0]$.

We can also consider another combination of information granules in which $\{A_i\}$ is a family of fuzzy sets and X is a set (see Figure 3.6). This leads us to the concept of rough fuzzy sets.

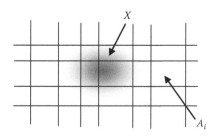

FIGURE 3.4
The development of the fuzzy rough set.

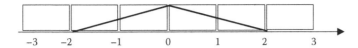

FIGURE 3.5
Family of generic descriptors, fuzzy set, and its representation in the form of some rough sets.

Alternatively, we can envision a situation in which both $\{A_i\}$ and X are fuzzy sets. The result comes with the lower and upper bound whose computing follows the formulas presented above.

3.3 Type-2 Fuzzy Sets

Type-2 fuzzy sets form an intuitively appealing generalization of interval-valued fuzzy sets. Instead of intervals of numeric values of membership degrees, we allow for the characterization of membership by fuzzy sets themselves. Consider a certain element of the universe of discourse, that is, x. The membership of x to A is captured by a certain fuzzy set formed over the unit interval. This construct generalizes the fundamental idea of a fuzzy set and helps relieve us from the restriction of having single numeric values describing a given fuzzy set (Mendel, 2001; Karnik, Mendel, Liang, 1999). An example of a type-2 fuzzy set is illustrated in Figure 3.7.

With regard to these forms of generalizations of fuzzy sets, there are two important facets that should be taken into consideration. First, there should be a clear motivation and a straightforward need to develop and use them. Second, it is imperative that there is sound membership determination procedure in place with which we can construct the pertinent fuzzy set.

To elaborate on these two issues, let us discuss a situation in which we deal with several databases populated by data coming from different regions of the same country. Using them we build a fuzzy set describing a concept of

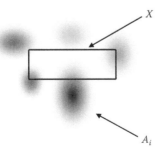

FIGURE 3.6
The concept of a rough fuzzy set.

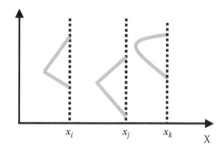

FIGURE 3.7
An illustration of type-2 fuzzy set; for each element of **X**, there is a corresponding fuzzy set of membership grades.

high income where the descriptor *high* is modeled as a certain fuzzy set. Given the experimental evidence, the principle of justifiable granularity could be a viable development alternative to pursue. Induced by some locally available data, the concept could exhibit some level of variability yet we may anticipate that all membership functions might be quite similar being reflective of some general commonalities. Given that the individual estimated membership functions are trapezoidal (or triangular). We can consider two alternatives to come up with some aggregation of the individual fuzzy sets (see Figure 3.8).

To fix the notation, let for "P" databases, the corresponding estimated trapezoidal membership functions be denoted by $A_1(x, a_1, m_1, n_1, b_1)$, $A_2(x, a_2, m_2, n_2, b_2)$,...., $A_P(x, a_P, m_P, n_P, b_P)$, respectively. The first aggregation alternative leads to the emergence of an interval-valued fuzzy set $A(x)$. Its membership function assumes interval values where for each "x" the interval of possible values of the membership grades is given in the form $[\min_i A_i(x, a_i, m_i, n_i, b_i), \max_i A_i(x, a_i, m_i, n_i, b_i)]$.

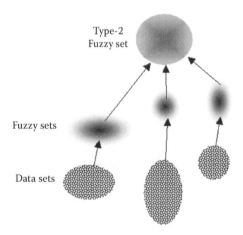

Type-2
Fuzzy set

Fuzzy sets

Data sets

FIGURE 3.8
A scheme of aggregation of fuzzy sets induced by "P" data sets.

The form of the interval-valued fuzzy set may be advantageous in further computing, yet the estimation process could be very conservative leading to very broad ranges of membership grades (which is particularly visible when dealing with different data and fuzzy sets induced on their basis).

3.4 Interval-Valued Fuzzy Sets

When defining or estimating membership functions or membership degrees, one may argue that characterizing membership degrees as single numeric values could be counterproductive and even somewhat counterintuitive given the nature of fuzzy sets themselves. Some remedy could be sought along the line of capturing the semantics of fuzzy sets through intervals of possible membership grades rather than single numeric entities (Hisdal, 1981; Gorzalczany, 1989). This gives rise to the concept of so-called interval-valued fuzzy sets. Formally, an interval-valued fuzzy set A is defined by two mappings from **X** to the unit interval $A = (A_-, A_+)$ where A_- and A_+ are the lower and upper bound of membership grades, $A_-(x)$ $A_+(x)$ for all $x \in$ **X** where $A_-(x) \leq A_+(x)$. The bounds are used to capture the effect of a lack of uniqueness of numeric membership—not knowing the detailed numeric values we admit that there are some bounds of possible membership grades. Hence, the name of the interval-valued fuzzy sets emerges, which becomes very much descriptive of the essence of the construct. The broader the range of the membership values, the less specific we are about membership degree of the element to the information granule. An illustration of the interval-valued fuzzy set is included in Figure 3.9. Again the principle of justifiable granularity can serve as a viable design vehicle.

In particular, when $A_-(x) = A_+(x)$, we end up with a *standard* (type-1) fuzzy set. The operations on interval-valued fuzzy sets are defined by considering separately the lower and upper bounds describing ranges of membership

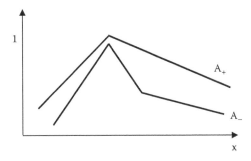

FIGURE 3.9
An illustration of an interval-valued fuzzy set. Note that the lower and upper bound of possible membership grades could differ quite substantially across the universe of discourse.

degrees. Given two interval-valued fuzzy sets A = (A_, A_+) and B = (B_, B_+) their union, intersection, and complement are introduced as follows,

Union (∪)   [min(A_(x), B_(x)), max (A_+(x), B_+(x))]

Intersection (∩)   [max(A_(x), B_(x)), min(A_+(x), B_+(x))] if A and B intersect

Complement   [1_A_+(x), 1_A_(x)]   (3.3)

3.5 Probabilistic Sets

The idea of probabilistic sets introduced by Hirota (1981) (see also Hirota and Pedrycz, 1984) has emerged in the context of pattern recognition where the existence of various sources of uncertainty is recognized including ambiguity of objects and their properties and the subjectivity of observers.

From the general conceptual perspective, probabilistic sets build upon fuzzy sets in which instead of numeric membership grades of the membership function, we consider associated probabilistic characteristics. Formally, a probabilistic set A on **X** is described by a defining function

$$A: \mathbf{X} \times \Omega \rightarrow \Omega_c \qquad (3.4)$$

where A is the (**B**, **B**_c)—measurable function for any x ∈ **X**. Here (Ω, **B**, P) is a parameter space, (Ω_c, **B**_c) = ([0,1], Borel sets) is a characteristic space, and {m| m: Ω × Ω_c (**B**, **B**_c) measurable function} is a family of characteristic variables.

From a formal point of view, a probabilistic set can be treated as a random field; to emphasize this, we use a notation A(x, ω).

With each x ∈ **X** one can associate a probability density function $p_x(u)$ defined over [0,1]. Alternatively, other probabilistic descriptors, for example, probability cumulative function $F_x(u)$, can be used.

Probabilistic sets come with an interesting moment analysis. For any fixed x, the mean value E(A(x)), variance V(A(x)), and the *n*-th moment denoted as $M^n(A(x))$ are considered

$$E(A(x)) = \int_0^1 p_x(u)\,du = M^1(A(x))$$

$$V(A(x)) = \int_0^1 (u - E(A(x))^2 p_x(u)\,du$$

$$M^n(A(x)) = \int_0^1 p_x^n(u)\,du \qquad (3.5)$$

where $p_x(u)$ is a probability density function for a given element of the universe (x) of discourse. An interesting relationship holds for the moments

$$M^1(A(x)) \geq M^2(A(x)) \geq \quad \geq M^3(A(x)) \geq \dots \tag{3.6}$$

which stresses that most of the information (description) about A is confined only to a few lowest moments. This also implies that higher-type information granules are not always justifiable and their practical relevance could be limited. Noticeably, the mean E(A) can be sought as a membership function used in fuzzy sets.

Assuming that the corresponding probabilistic characteristics are available, we can determine results for logic operations on probabilistic sets (Czogala and Pedrycz, 1983; Czogala, Gottwald, and Pedrycz, 1983). For instance, let us determine the result for the union and intersection of two probabilistic sets A and B. For the cumulative probability functions of A and B, $F_{A(x)}$ and $F_{B(x)}$, we obtain

$$F_{\max(A,B)(x)}(u) = F_{A(x)}(u)F_{B(x)}(u) \tag{3.7}$$

and

$$F_{\min(A,B)(x)}(u) = F_{A(x)}(u) + F_{B(x)}(u) - F_{A(x)}(u)F_{B(x)}(u) \tag{3.8}$$

where $u \in [0,1]$.

3.6 Hybrid Models of Information Granules: Probabilistic and Fuzzy Set Information Granules

In a number of real-world situations, we describe concepts and form models using several formalisms of information granularity. These constructs become of particular interest when information granules have to capture a multifaceted nature of the problem. There are a number of interesting application-driven scenarios. The one that deserves attention here concerns a model involving probability and fuzzy sets. From the very inception of fuzzy sets, there were quite vivid debates about linkages between these two formal vehicles of information granulation. There was a claim that these two formalisms are the same, a prevailing and justifiable position is that probability and fuzzy sets are orthogonal by capturing two very distinct facets of reality. Let us stress that when dealing with probabilities, we are concerned with an occurrence (or nonoccurrence) of a certain event (typically described by some set). Fuzzy sets are not about occurrence but about a degree of membership to a certain concept. In the early studies by

Zadeh (1968), the aspect of orthogonality was raised very clearly with the notions such as fuzzy events, probability of fuzzy events, and the like. It is instructive to recall the main definitions introduced in the literature (Zadeh, 1968). As usual, the probability space is viewed as a triple $(\mathbf{R}^n, B, P)$ where B is the σ-field of Borel sets in $\mathbf{R}^n$ and P is a probability measure over $\mathbf{R}^n$. If $A \in B$ then the probability of A, P(A) can be expressed as $P(A) = \int_{\mathbf{R}^n} A(\mathbf{x})dP$ or using the characteristic function of A, one has $P(A) = \int_{\mathbf{R}^n} A(\mathbf{x})dP = E(A)$. In other words, P(A) is an expected value of the characteristic function of A.

The generalization comes in the form of a so-called fuzzy event A where A is a fuzzy set in $\mathbf{R}^n$ with a Borel measurable membership function.

The probability of a fuzzy event A is the Lebesgue–Stieltjes integral

$$P(A) = \int_{\mathbf{R}^n} A(\mathbf{x})dP = E(A) \tag{3.9}$$

The mean (expected value) of the fuzzy event A is

$$m(A) = \int_{\mathbf{R}^n} \mathbf{x}A(\mathbf{x})dP \tag{3.10}$$

The variance of A defined in $\mathbf{R}$ is given in the form

$$Var(A) = \int_{\mathbf{R}} (x - A(x))^2\, dP \tag{3.11}$$

The generalization of the entropy of A defined in the finite space $\{x_1, x_2, \ldots, x_n\}$ with probabilities $p_1, p_2, \ldots, p_n$ comes in the form (Zadeh, 1968)

$$H(A) = -\sum_{i=1}^{n} A(x_i)p_i \log(p_i) \tag{3.12}$$

It is quite intuitive that probability of a fuzzy event and its specificity are interrelated. Considering that the specificity measure characterizes how detailed (specific) a certain piece of knowledge A is, it can be regarded as a decreasing function of a sigma count of the fuzzy event (assuming that A is a normal fuzzy set) such that $Spec(A) = 1$ for $A = \{x0\}$ and $Spec(\mathbf{X}) = 0$ where $\mathbf{X}$ is the universe of discourse (space) over which information granules are formed. Along with the specificity, which we wish to see as high as possible (so that the piece of knowledge conveyed by A is highly meaningful), we also require that the probability of the fuzzy event is high enough. Imposing some threshold values on these two requirements, that is, λ_1 for

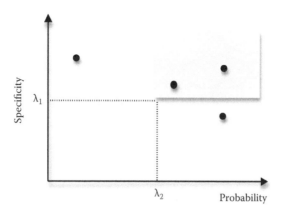

FIGURE 3.10
Information granules described in terms of their specificity and probability. Shown in a feasible region (shaded area) and visualized are several information granules satisfying and not satisfying the requirements identified by the thresholds.

the specificity, and λ_2 for the probability, we express the above requirements as follows: Spec(A) > λ_1 and Prob(A) > λ_2 (see Figure 3.10) showing a region of feasible information granules. In this way, given an information granule A, one can easily check if it is feasible with regard to the two requirements presented above. Furthermore, one can think of possible adjustments of the information granule by decreasing its specificity in order to increase the probability or moving it around the universe of discourse to increase specificity and probability value.

There are also some other interesting and practically viable linguistic descriptions in which we encounter so-called linguistic probabilities (Zadeh, 2002) such as *low* probability, *high* probability, probability *about* 0.7, *very likely*, and so forth, that are applied to describe events or fuzzy events or events described by any other information granule.

The pieces of knowledge such as "it is *very high* probability that inflation rate will increase *sharply*" or "*high* error value occurs with a *very low* probability" are compelling examples of granular descriptors built with the use of probabilistic and fuzzy set information granules.

3.7 Realization of Fuzzy Models with Information Granules of Higher Type and Higher Order

In a nutshell, fuzzy models can be viewed as models built on the basis of information granules. For instance, fuzzy rule-based models revolve around a collection of rules. Both the condition and conclusion parts of the

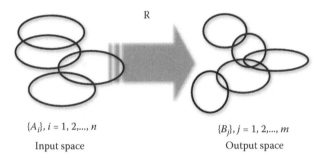

$\{A_i\}, i = 1, 2,..., n$ $\{B_j\}, j = 1, 2,..., m$

Input space   Output space

FIGURE 3.11
Fuzzy model as a web of connections between information granules.

rules are information granules. In Takagi–Sugeno rules, the condition parts of the rules are information granules while the conclusion parts are functional models (local regression models). Fuzzy models form relationships among information granules in the input and output space (see Figure 3.11).

The web of relationships can be expressed in many different ways. For instance, a certain fuzzy relation R can be envisioned with association to some composition operator * using which one realizes the following mapping

$$y = z * R \qquad (3.13)$$

where z is a vector of activation levels of information granules $A_1, A_2, ..., A_n$ by a given input x whereas y is a vector of activation of information granules $B_1, B_2, ..., B_m$ positioned in the output space, dim $(y) = m$. Typically, the composition operator (*) is realized as a max-t composition or s-t composition involving t-norms and t-conorms. In this case, Equation (3.13) is regarded as a fuzzy relational equation. A careful look at this relationship allows us to conclude that the fuzzy model is created at the level of information granules so both z and y are fuzzy sets of order 2 (as being formed in the space of information granules). In brief, Equation (3.13) is a fuzzy model of order 2.

The construction of the fuzzy relation R is supported by the theory of fuzzy relational equations including approximate methods aimed at the determination of their solutions. The fuzzy relation is estimated in the presence of learning data and comes through the minimization of a certain performance index. The entries of the fuzzy relation are numeric, that is, $R = [r_{ij}]$, where rij are located in [0,1]. The relational model is not ideal and its quality can be evaluated by forming a granular relation, $G(R)$ (where the entries of $G(R)$ are information granules, that is, intervals positioned in the unit interval) by realizing an optimal allocation of information (see Figure 3.12). In this way, we come up with an interesting and useful augmentation of the existing fuzzy model by forming a *granular* fuzzy model of order 2.

There might be a different identification scenario in which for the same family of information granules, on a basis of different data sets, one realizes

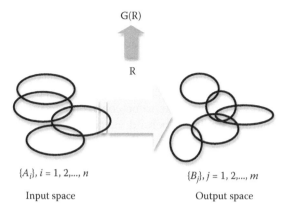

FIGURE 3.12
From fuzzy model of order 2 to granular fuzzy model of order 2.

a family of fuzzy models of order 2 (see Figure 3.13). Then the individual models are reconciled by forming a global (holistic) view of the system. The reconciliation of the individual models is possible by making the model at the higher level of hierarchy granular. Again, we arrive at the granular fuzzy model of order 2. Its development is realized by means of the principle of justifiable granularity.

3.8 Conclusions

A wealth of formal frameworks of information granules is essential in the description and handling of a diversity of real-world phenomena. We stressed the need behind the formation of information granules of higher

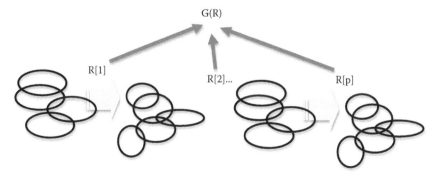

FIGURE 3.13
Reconciliation of local fuzzy models by forming a granular fuzzy model at the higher level of hierarchy.

order and higher type. When arranged together they are helpful to cope with the complexity of the descriptors. One has to be aware that although conceptually it is possible to develop a multilevel hierarchy of high order or/and high type of information granules, pursuing an excessive number of levels of the hierarchy could be counterproductive. Of course, one may argue that type-2 fuzzy sets could be extended (generalized) to type-3, type-4, and so forth, fuzzy sets; however, one has to be cognizant of the need to provide strong and compelling evidence behind successive levels of the hierarchy and the associated estimation costs of the determination of such hierarchically structured information granules.

The hybrid constructs form an interesting avenue to investigate as they invoke a slew of synergistic constructs (see, for example, Zhang, 2012). While we showed the probabilistic–linguistic synergy, some other alternatives might be worth pursuing.

References

Czogala, E., S. Gottwald, and W. Pedrycz. 1983. Logical connectives of probabilistic sets. *Fuzzy Sets and Systems*, 10, 299–308.

Czogala, E. and W. Pedrycz. 1983. On the concept of fuzzy probabilistic controllers. *Fuzzy Sets and Systems*, 10, 109–121.

Gorzałczany, M.B. 1989. An interval-valued fuzzy inference method—Some basic properties. *Fuzzy Sets and Systems*, 31, 243–251.

Hirota, K. 1981. Concepts of probabilistic sets. *Fuzzy Sets and Systems*, 5, 1, 31–46.

Hirota, K. and W. Pedrycz. 1984. Characterization of fuzzy clustering algorithms in terms of entropy of probabilistic sets. *Pattern Recognition Letters*, 2, 4, 213–216.

Hisdal, E. 1981. The IF THEN ELSE statement and interval-valued fuzzy sets of higher type. *International Journal of Man-Machine Studies*, 15, 385–455.

Karnik, N.M., J.M. Mendel, and Q. Liang. 1999. Type-2 fuzzy logic systems. *IEEE Transactions on Fuzzy Systems*, 7, 643–658.

Mendel, J. 2001. *Uncertain Rule-Based Fuzzy Logic Systems: Introduction and New Directions*. Upper Saddle River, NJ: Prentice Hall.

Zadeh, L.A. 1968. Probability measures of fuzzy events. *J. Math. Anal. Appl.*, 23, 421–427.

Zadeh, L.A. 2002. Toward a perception-based theory of probabilistic reasoning with imprecise probabilities. *J. Statistical Planning and Inference*, 105, 233–64.

Zhang, Z. Forthcoming 2012. On characterization of generalized interval type-2 fuzzy rough sets. *Information Sciences*.

4

Representation of Information Granules

Any numeric entity or information granule can be represented (described) by a finite family of information granules. We emphasize the nature of the representation in terms of processes of granulation and degranulation. It becomes interesting to look at how such a description can be realized and which of its parameters may impact the quality of the resulting representation. We also show that the representation of this form is inherently associated with the emergence of higher-type information granules. Some linkages with the construction of vector quantization and analog-to-digital and digital-to-analog conversion are investigated.

4.1 Description of Information Granules by a Certain Vocabulary of Information Granules

Given is a certain finite vocabulary of information granules $\mathbf{A} = \{A_1, A_2, \ldots, A_c\}$ where A_i is a certain information granule expressed in the framework of any of the formalisms presented before. Both X and the elements of $\mathbf{A}$ are defined in the same universe of discourse $\mathbf{X}$. We are interested in expressing (describing) a new evidence X (either numeric or granular) in terms of the elements of $\mathbf{A}$. To assure the soundness of $\mathbf{A}$ (its ability to describe X), we assume that $\mathbf{A}$ is a partition of $\mathbf{X}$, that is, the condition $A_1(x) + A_2(x) + \ldots + A_c(x) = 1$ for any $x \in \mathbf{X}$. This requirement could be made less restrictive by requesting that each element in $\mathbf{X}$ belongs to at least one A_i with a nonzero membership level, that is, $\forall_x \exists_i A_i(x) > 0$.

Intuitively, one could envision a number of possible ways of describing X in terms of the elements of this vocabulary. A way that is quite appealing is to describe a relationship between X and A_i in terms of coincidence (overlap) of these two and an inclusion of A_i in the information granule of interest (see Figure 4.1).

The degree of overlap (intersection) known also as a possibility measure of X with A_i is computed

$$\text{Poss}(X, A_i) = \sup_{x \in X}[X(x)tA_i(x)] \tag{4.1}$$

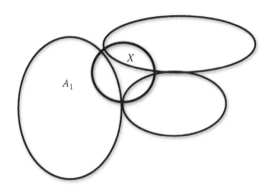

FIGURE 4.1
A collection of elements of **A** and information granule X.

The degree of inclusion of A_i in X is determined in the following way:

$$\text{Nec}(X, A_i) = \inf_{x \in X}[(1 - A_i(x))sX(x)] \tag{4.2}$$

where "t" and "s" are some t-norms and t-conorms, respectively [11]. In virtue of Equations (4.1) and (4.2), the following relationship holds $\text{Nec}(X, A_i) \leq \text{Poss}(X, A_i)$. If the information granule X is in a degenerated form, that is, it comes as a single numeric entity, $A(x) = \delta(x - x_0)$, namely,

$$A(x) = \begin{cases} 1 & \text{if } x = x_0 \\ 0 & \text{otherwise} \end{cases} \tag{4.3}$$

then the degree of overlap coincides with the degree of inclusion; one has $\text{Poss}(X, A_i) = \text{Nec}(X, A_i) = A(x_0)$. In case X is a fuzzy set with a continuous membership function and A_i are intervals $A_i = [a_i, a_{i+1}]$, then the possibility and necessity degrees are the maximal and minimal value of X achieved in the interval $[a_i, a_{i+1}]$, that is, $\text{Poss}(X, A_i) = \max_{x \in [a_i, a_{i+1}]} X(x)$ and $\text{Nec}(X, A_i) = \min_{x \in [a_i, a_{i+1}]} X(x)$. The plot shown in Figure 4.2 highlights the underlying idea.

It is worth noting that the concepts of overlap and inclusion (no matter how they are realized) are intuitively appealing in the realization of relationships between two information granules. In particular, they are used in the concept of rough sets to define lower and upper approximations.

Having determined the possibility and necessity values for each element of **A**, we can formulate a reconstruction problem: determine X given the vocabulary **A** and the levels of inclusions and overlap calculated for each of the elements of **A**.

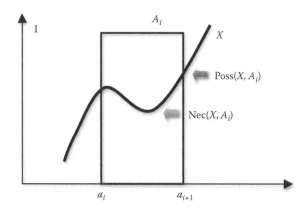

FIGURE 4.2
Computing the overall and inclusion degrees for the interval information granules of **A**.

By taking advantage of the fundamental results of fuzzy relational equations (Di Nola et al., 1989; Pedrycz, 1989; Hirota and Pedrycz, 1999; Nobuhara, Pedrycz, and Hirota, 2000) (as both Equations 4.1 and 4.2 can be viewed as sup-t and inf-s fuzzy relational equations to be solved with respect to X), we obtain the bounds of the reconstruction process for Equation (4.1); the upper bound is

$$\hat{X}(x) = \min_{i=1,2,\ldots,c} [A_i(x) \phi Poss(X, A_i)] \tag{4.4}$$

for Equation (4.2), the lower bound is

$$\check{X}(x) = \max_{i=1,2,\ldots,c} [(1 - A_i(x)) \beta Nec(X, A_i)] \tag{4.5}$$

The two aggregation operators are defined as follows:

$$a\phi b = \sup\{c \in [0,1] \,|\, atc \le b\} \quad a\beta b = \inf\{c \in [0,1] \,|\, asc \ge b\} \quad a, b \in [0,1] \tag{4.6}$$

Several selected examples of these operators associated with the corresponding t-norms and t-conorms are covered in Table 4.1.

Interestingly, the reconstruction of X is no longer unique. Instead, we obtain the bounds of the reconstruction with the lower and upper bounds expressed by Equations (4.5) and (4.6), respectively. In other words, the original information granule X is reconstructed in the form of a certain information granule of higher type. In other words, we note that there is an effect of the elevation of the granularity level of the result when dealing with the representation and the ensuing reconstruction problem. Continuing with

TABLE 4.1

Selected Examples of ϕ and β Operators; $a, b \in [0,1]$

T-norm	ϕ operator	β operator
Godel, min(a, b)	$a\phi b = \begin{cases} b & \text{if } a>b \\ 1, & \text{otherwise} \end{cases}$	$a\beta b = \begin{cases} b & \text{if } a<b \\ 0, & \text{otherwise} \end{cases}$
Product ab	$a\phi b = \begin{cases} b/a & \text{if } a>b \\ 1, & \text{otherwise} \end{cases}$	$a\beta b = \begin{cases} \dfrac{b-a}{1-a} & \text{if } a<b \\ 0, & \text{otherwise} \end{cases}$
Lukasiewicz max(0, a+b−1)	$a\phi b = \begin{cases} 1-a+b & \text{if } a>b \\ 1, & \text{otherwise} \end{cases}$	$a\beta b = \begin{cases} b-a & \text{if } a<b \\ 0, & \text{otherwise} \end{cases}$

Note: T-conorm is dual to the t-norm shown in the table.

the example of interval information granules forming the vocabulary, we obtain the results:

$$\hat{X}_i(x) = \begin{cases} 1, & \text{if } A_i(x)=0 \\ A_i(x)\phi\text{Poss}(X, A_i), & \text{if } A_i(x)=1 \end{cases} = \begin{cases} 1, & \text{if } A_i(x)=0 \\ \text{Poss}(X, A_i), & \text{if } A_i(x)=1 \end{cases}$$

$$= \begin{cases} 1, & \text{if } x \notin [a_i, a_{i+1}] \\ \text{Poss}(X, A_i), & \text{if } x \in [a_i, a_{i+1}] \end{cases} \tag{4.7}$$

and

$$\check{X}_i(x) = \begin{cases} 0, & \text{if } A_i(x)=0 \\ (1-A_i(x))\beta\text{Nec}(X, A_i), & \text{if } A_i(x)=1 \end{cases} = \begin{cases} 0, & \text{if } A_i(x)=0 \\ \text{Nec}(X, A_i), & \text{if } A_i(x)=1 \end{cases}$$

$$= \begin{cases} 0, & \text{if } x \notin [a_i, a_{i+1}] \\ \text{Nec}(X, A_i), & \text{if } x \in [a_i, a_{i+1}] \end{cases} \tag{4.8}$$

(see also Figure 4.3). It becomes apparent that the result is an interval-valued construct with the bounds determined by the maximal and minimal value of X taken over $[a_i, a_{i+1}]$.

Applying the above algorithm to the reconstruction (degranulation) completed in the scenario portrayed in Figure 4.4, the result becomes a shadowed set with several regions of shadows resulting from the bounds formed by Equations (4.4) and (4.5). Compare this result with the definition of the rough set—those two concepts coincide.

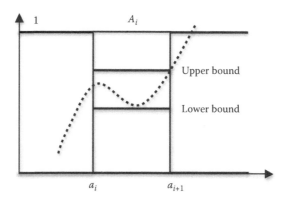

FIGURE 4.3
Result of reconstruction realized with the use of A_is.

In the case of a multidimensional problem where A_i and X are in the form of the Cartesian products of their components defined in the corresponding universes of discourse, $A_i = A_{i1} \times A_{i2} \times \ldots \times A_{ip}$ and $X = X_1 \times X_2 \times \ldots \times X_p$, the formulas for the overlap and inclusion as well as the reconstruction expressions are as follows:

$$Poss(X, A_i) = \sup_{x \in X}[X(x)tA_i(x)] =$$

$$\sup_{x_1, x_2 \ldots x_p}[\min(X(x_1)tA_{i1}(x_1), X(x_2)tA_{i2}(x_2), \ldots, X(x_p)tA_{ip}(x_p)] \quad (4.9)$$

$$Nec(X, A_i) = \inf_{x \in X}[(1 - A_i(x))sX(x)] \quad (4.10)$$

where $x = [x_1 \ x_2 \ \ldots \ x_n]$. In virtue of the fundamental results of the theory of fuzzy relational equations (and Boolean equations), one can show that X is always positioned in-between the lower and upper bound. This poses several interesting and practically relevant optimization problems:

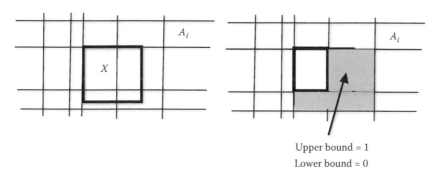

FIGURE 4.4
Granulation and degranulation leading to the shadowed sets.

Optimization of the vocabulary of granular terms. For the predetermined number of terms (c), distribute them in such a way that the reconstruction bounds are made as tight as possible (the distance $||.||$ is the lowest one). Formally, the problem is posed in the form

$$\text{Min}_A \sum_{X \in F} \| \hat{X} - \breve{X} \| \tag{4.11}$$

where the granulation and degranulation are realized for all Xs coming from a certain collection of information granules **F**. The problem itself can be structured in several ways. For instance, assuming the form of the information granules (i.e., Gaussian membership functions), one optimizes the parameters so that the difference between the bounds is minimized. The more advanced formulation of the problem could involve a selection of the form of the information granules (in the case of fuzzy sets, we can talk about triangular, trapezoidal, S-shape, and Gaussian membership functions); this structure-oriented aspect of the optimization has to be considered. Typically, population-based optimization tools are highly suitable here.

Optimization of the dimensionality of the vocabulary of granular terms. This problem is a generalization of the one outlined above in the sense that we allocate different numbers of information granules (vocabulary) across the individual coordinates of the multivariable space. In other words, the number of information granules in **A** being explicitly formed in the coordinates of **X** could vary among the coordinates. Assuming that the overall size of the vocabulary is predetermined, that is, equal to "c," the optimization involves an allocation of the number of the information granules along the coordinates so that the sum $c_1 + c_2 + \dots + c_n$ is kept equal to "c."

Formation of hierarchical granulation structure. While the number of information granules of **A** could be limited, a hierarchical structure in which some of these granules are further *expanded* by forming a collection of more detailed, refined granules and thus enhancing the quality of the degranulation process.

4.2 Information Granulation–Degranulation Mechanism in the Presence of Numeric Data

When dealing with any numeric data, we are concerned with their representation in terms of a collection of information granules. More formally, we can describe this representation process as a way of expressing any data point x in terms of the information granules and describe the result as a vector in the c-dimensional hypercube, namely $[0, 1]^c$,

$$G: \mathbf{R}^n \to [0, 1]^c \qquad (4.12)$$

The degranulation step is about a reconstruction of $\mathbf{x}$ on the basis of the family of information granules (clusters). It can be treated as a certain mapping

$$G^{-1}: [0, 1]^c \to \mathbf{R}^n \qquad (4.13)$$

The capabilities of the information granules to reflect the structure of the original data can be conveniently expressed by comparing how much the result of degranulation, that is, $\mathbf{x}$, differs from the original pattern $\mathbf{x}$, that is, $\hat{\mathbf{x}} \neq \mathbf{x}$. More formally, $\hat{\mathbf{x}} = G^{-1}(G(\mathbf{x}))$ where G and G^{-1} denote the corresponding phases of information granulation and degranulation (Pedrycz and Oliveira, 1996, 2008).

The crux of the granulation–degranulation principle is visualized in Figure 4.5. Note the transformations G and G^{-1} operate between the spaces of data and information granules.

Let us start with the granulation phase. More specifically, $\mathbf{x}$ is expressed in the form of the membership grades u_i of $\mathbf{x}$ to the individual granules A_i, which form a solution to the following optimization problem:

$$\text{Min} \sum_{i=1}^{c} u_i^m(\mathbf{x}) \|\mathbf{x} - \mathbf{v}_i\|^2 \qquad (4.14)$$

subject to the constraints imposed on the degrees of membership

$$\sum_{i=1}^{c} u_i(\mathbf{x}) = 1 \quad u_i(\mathbf{x}) \in [0,1] \qquad (4.15)$$

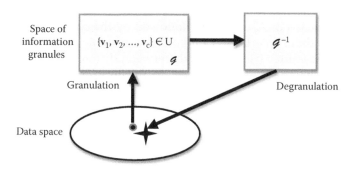

FIGURE 4.5
The granulation–degranulation mechanism as a realization of mapping between data space and the space of information granules.

where "m" stands for the so-called fuzzification coefficient assuming values greater than 1, m > 1 [6]. From now on assume that the distance function is the Euclidean one. The derived solution to the problem (for the details refer also to Chapter 5) comes in the form,

$$u_i(x) = \frac{1}{\sum\limits_{j=1}^{c} \left(\dfrac{\| x - v_i \|}{\| x - v_j \|} \right)^{2/(m-1)}}$$

(4.16)

For the degranulation phase, given $u_i(x)$ and the prototypes v_i, the vector is considered as a solution to the minimization problem in which we reconstruct (degranulate) original x when using the prototypes and the membership grades

$$\sum_{i=1}^{c} u_i^m(x) \| \hat{x} - v_i \|^2$$

(4.17)

Because of the use of the Euclidean distance in the above performance index, the calculations here are straightforward yielding the result

$$\hat{x} = \frac{\sum\limits_{i=1}^{c} u_i^m(x) v_i}{\sum\limits_{i=1}^{c} u_i^m(x)}$$

(4.18)

It is important to note that the description of x in more abstract fashion realized by means of A_i and being followed by the consecutive degranulation brings about a certain granulation error (which is inevitable given the fact that we move back and forth between different levels of abstraction). While the above formulas pertain to the granulation realized by fuzzy sets, the granulation–degranulation error is also present when dealing with sets (intervals). In this case we are faced with a quantization error, which becomes inevitable when working with A/D (granulation) and D/A (degranulation) conversion mechanisms. The problem formulated above is also associated with vector quantization, which has been widely studied in the literature, cf. Linde, Buzo, and Gray (1988), Gersho and Gray (1992), Gray (1984), Yair, Zeger, and Gersho (1992), and Lendasse et al. (2005).

The quality of the granulation–degranulation scheme depends upon a number of parameters and they can be optimized. The performance index considered here comes in the form

$$V = \sum_{x_k \in D} \| x_k - \hat{x}_k \|^2 \qquad (4.19)$$

with x_k coming from some set **D**. The two of them are the most visible: (a) the number of information granules in **A**, and (b) fuzzification coefficient (m). The role of the first one is apparent: the increase of the cardinality of A leads to the reduction of the reconstruction error. The second parameter changes the form of the associated membership; the values close to 1, m→1 lead to the membership functions close to the characteristic functions present in set theory. Its optimal value is problem dependent. A series of typical curves presenting the dependence of the reconstruction error is visualized in Figure 4.6; see Pedrycz and Oliveira (2008) for extensive experimental studies. The larger the codebook is, the smaller the reconstruction error becomes. The optimal value of "m," however, is data dependent. Interestingly, in light of the criterion discussed here, the choice m = 2, which is commonly encountered in the literature is not always justified; the optimal values of the fuzzification can vary in a quite broad range of values.

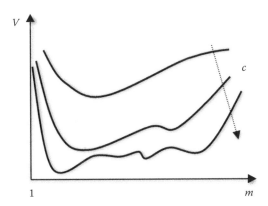

FIGURE 4.6
Reconstruction error versus the fuzzification coefficient for selected values of the size of the codebook **A**.

4.3 Granulation–Degranulation in the Presence of Triangular Fuzzy Sets

An interesting result showing a lossless degranulation (reconstruction) can be achieved in case x are real numbers. Interestingly, the results in the case of one-dimensional and vector quantization are well reported in the literature, cf. Gersho and Gray (1992), Gray (1984), Linde, Buzo, and Gray (1988), and Yair, Zeger, and Gersho (1992). While the quantization error could be minimized, the method is in essence lossy (coming with a nozero quantization error). In contrast, a codebook formed by fuzzy sets can lead to the zero error of the quantization. With this regard we show a surprisingly simple yet powerful result whose essence could be summarized as follows: a codebook formed of fuzzy triangular fuzzy sets (fuzzy numbers) with an overlap of one half between two neighboring elements (see Figure 4.7) leads to the lossless reconstruction mechanism. The essence of the scheme can be formulated in the form of the following proposition.

Proposition

Let us assume the following:

(a) The fuzzy sets of the codebook $\{A_i\}$, $i = 1, 2, \ldots, c$ form a fuzzy partition, $\sum_{i=1}^{c} A_i(x) = 1$, and for each x in **X** at least one element of the codebook is *activated*, that is, $A_i(x) > 0$

(b) For each x, only two neighboring elements of the codebook are *activated*, that is, $A_1(x) = 0, \ldots, A_{i-1}(x) = 0, A_i(x) > 0, A_{i+1}(x) > 0, A_{i+2}(x) = \ldots = A_c(x) = 0$

(c) The decoding is realized as a weighted sum of the activation levels and the prototypes of the fuzzy sets v_i, namely, $\hat{x} = \sum_{i=1}^{c} A_i(x)v_i$

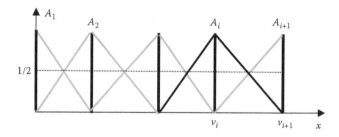

FIGURE 4.7
An example of the codebook composed of triangular fuzzy sets with an overlap of one half between each two neighboring elements of the codebook.

Then the elements of the codebook described by piecewise linear membership functions

$$A_i(x) = \begin{cases} \dfrac{x - v_{i-1}}{v_i - v_{i-1}} & \text{if } x \in [v_{i-1}, v_i] \\[2ex] \dfrac{x - v_{i+1}}{v_i - v_{i+1}} & \text{if } x \in [v_i, v_{i+1}] \end{cases} \tag{4.20}$$

lead to the zero decoding error (lossless compression) meaning that $\hat{x} = x$. ∎

Proof: To show the proof, we consider any element x lying in the interval $[v_i, v_{i+1}]$. In virtue of the properties of $\{A_i\}$ presented above, we rewrite the decoding formula $\hat{x} = \sum_{i=1}^{c} A_i(x)v_i$ as follows:

$$\hat{x} = A_i(x)v_i + (1 - A_i(x))v_{i+1} \tag{4.21}$$

We request a lossless compression meaning that $\hat{x} = x$. In other words, it is required that

$$x = A_i(x)v_i + (1 - A_i(x))v_{i+1} \tag{4.22}$$

The formula for $A_i(x)$ (which describes the right-hand side of the membership function of A_i and spreads in-between v_i and v_{i+1}) is obtained by rearranging the terms in Equation (4.22). This leads to the expression

$$A_i(x)\,(v_i - v_{i+1}) = x - v_{i+1}$$

and

$$A_i(x) = \frac{x - v_{i+1}}{v_i - v_{i+1}} \tag{4.23}$$

The above points at the piecewise linear character of the membership function of A_i. In the same fashion, we can deal with the left-hand side of the membership function of A_i by considering the interval $[v_{i-1}, v_i]$. In this case, we can demonstrate that for all x in $[v_{i-1}, v_i]$ the membership comes in the form $A_i(x) = \frac{x - v_{i-1}}{v_i - v_{i-1}}$ which completes the proof.

Interestingly enough, triangular fuzzy sets have been commonly used in the development of various fuzzy set constructs as will be discussed (models, controllers, classifiers, etc.), yet the lossless character of such codebooks is not generally known and utilized, perhaps with very few exceptions, cf. Pedrycz (1994).

The result obtained from the lossless information granulation with the use of fuzzy sets stands in a sharp contrast with the lossy granulation–degranulation realized with sets (intervals). In essence, in this case we encounter well-known schemes of analog-to-digital (A/D) conversion (granulation) and digital-to-analog (D/A) conversion (degranulation). It is obvious that such conversions produce an inevitable error.

4.4 Conclusions

With regard to encoding and decoding abilities, it is shown that any representation of information granules and numeric entities is associated with degranulation error, which could be absorbed by forming an information granule of higher type. For instance, while the original information granules are fuzzy sets, the results of degranulation become fuzzy sets of type 2 (interval-valued fuzzy sets, in particular). The optimization of the granular landmarks is a well-formulated problem guided by making the bounds of the interval as tight as possible. As a result, the codebook of the granular landmarks becomes instrumental in the discovery of the essential relationships in the collection of numeric or granular data.

The granulation and degranulation mechanisms are also essential in system modeling realized with the aid of information granules. The granulation error has to be taken into consideration when assessing the quality of models realized in such a framework. With this regard, in the one-dimensional case, the role of triangular fuzzy sets becomes highly advantageous with its delivery of the lossless granulation–degranulation scheme.

References

Di Nola, A., S. Sessa, W. Pedrycz, and E. Sanchez. 1989. *Fuzzy Relational Equations and Their Applications in Knowledge Engineering*. Dordrecht: Kluwer Academic Publishers.

Gersho, A. and R.M. Gray. 1992. *Vector Quantization and Signal Compression*. Boston, MA: Kluwer Academic Publishers.

Gray, R. 1984. Vector quantization. *IEEE Acoust. Speech, Signal Process. Mag.*, 1, 4–29.

Hirota, K. and W. Pedrycz. 1999. Fuzzy relational compression. *IEEE Trans. on Systems, Man, and Cybernetics*, 29, 3, 407–415.

Klement, E., R. Mesiar, and E. Pap. 2000. *Triangular Norms*. Dordrecht: Kluwer Academic Publishers.

Lendasse, A., D. Francois, V. Wertz, and M. Verleysen. 2005. Vector quantization: A weighted version for time-series forecasting. *Future Generation Computer Systems*, 21, 1056–1067.

Linde, Y., A. Buzo, and R. Gray. 1988. An algorithm for vector quantizer design. *IEEE Trans. Commun.* COM-28, 1, 84–95.

Nobuhara, H., W. Pedrycz, and K. Hirota. 2000. Fast solving method of fuzzy relational equation and its application to lossy image compression/reconstruction. *IEEE Trans. on Fuzzy Systems*, 8, 3, 325–335.

Pedrycz, W. (1989) 1993. *Fuzzy Control and Fuzzy Systems*, 3rd ext. ed. Taunton, New York: Research Studies Press/J. Wiley.

Pedrycz, W. 1994. Why triangular membership functions? *Fuzzy Sets & Systems*, 64, 21–30.

Pedrycz, W. and J. Valente de Oliveira. 1996. Optimization of fuzzy models. *IEEE Transaction on Systems, Man, and Cybernetics, Part B*, 26, 4, 627–636.

Pedrycz, W. and J. Valente de Oliveira. 2008. A development of fuzzy encoding and decoding through fuzzy clustering. *IEEE Transactions on Instrumentation and Measurement*, 57, 4, 829–837.

Yair, E., K. Zeger, and A. Gersho. 1992. Competitive learning and soft competition for vector quantizer design. *IEEE Trans. Signal Processing*, 40, 2, 294–309.

5

The Design of Information Granules

Information granules are building blocks reflective of domain knowledge about a problem. As such they are constructed on the basis of a variety of sources of experimental evidence. These pieces of evidence could come in the form of numeric data as well as perceptions (information granules) originating from experts. The existing evidence has to be structured into some tangible, meaningful, and easily interpretable information granules.

In this chapter, several main approaches to the design of information granules are presented. We start with the principle of justifiable granularity to show how a single information granule is constructed in the presence of available experimental evidence. The principle is general in the sense that both the form of the data over which an information granule is developed as well as the character of the resulting granule can vary across the formal frameworks not being confined to any of the particular formalism of information granularity. In the sequel, we revisit methods of clustering, especially fuzzy clustering, as a vehicle to construct information granules and look at the augmentation of clustering techniques (so-called knowledge-based clustering) so that they result in information granules built on the basis of numeric data as well as knowledge tidbits of human expertise which are some information granules themselves. Some refinements of information granules arising through their successive specialization are discussed. We present the idea of collaborative clustering in which higher-type information granules are formed on the basis of several sources of data.

5.1 The Principle of Justifiable Granularity

We are concerned with the development of a single information granule Ω based on some experimental evidence (data) coming in the form of a collection of a one-dimensional (scalar) numeric data, $\mathbf{D} = \{x_1, x_2, ..., x_N\}$. The essence of the principle of justifiable granularity is to form a meaningful (legitimate) information granule based on available experimental evidence (data) $\mathbf{D}$ where we require that such a construct has to adhere to the two intuitively compelling requirements:

(i) *Experimental evidence (legitimacy)*. The numeric evidence accumulated within the bounds of Ω has to be as *high* as possible. By requesting this, we anticipate that the existence of the information granule is well motivated (justified) as being reflective of the existing experimental data. For instance, if Ω is a set, then the more data included within the bounds of Ω, the better—in this way the set becomes more legitimate. Likewise, in the case of a fuzzy set, the higher the sum of membership degrees (σ-count) of the data in Ω is, the higher the justifiability of this fuzzy set.

(ii) *Sound semantics (meaning)*. At the same time, the information granule should be as *specific* as possible. This request implies that the resulting information granule comes with well-defined semantics (meaning). In other words, we would like to have Ω highly detailed, which makes the information granule semantically meaningful (sound). This implies that the smaller (more compact) the information granule (higher information granularity) is, the better. This point of view is in agreement with our general perception of knowledge being articulated through constraints (information granules) specified in terms of statements such as "x is A," "y is B," and so forth, where A and B are constraints quantifying knowledge about the corresponding variables. Evidently, the piece of knowledge coming in the form "x is in [1,3]" is more specific (semantically sound, more supportive of any further action, etc.) than another piece of knowledge where we know only that "x is in [0, 10]."

While these two requirements are appealing from an intuitive point of view, they have to be translated into some operational framework in which the formation of the information granule can be realized. This framework depends upon the accepted formalism of information granulation, namely, a way in which information granules are described as sets, fuzzy sets, shadowed sets, rough sets, probabilistic granules, and others.

For the clarity of presentation and to focus on the nature of the construct, let us start with an interval (set) representation of Ω. The requirement of experimental evidence is quantified by counting the number of data falling within the bounds of Ω. If **D** can be regarded as the samples drawn from a certain probability density function (pdf) $p(x)$, then the level of accumulated experimental evidence is expressed as the integral (cumulative probability) $\int_\Omega p(x)dx$. If the distribution of data is not available but a finite data set **D** is provided, we determine cardinality of elements falling within the bounds of Ω, namely card $\{x_k \,|x_k{\in}\Omega\}$. More generally, we may consider an increasing function (f_1) of this cardinality, that is, $f_1(\text{card}\{x_k \mid x_k \in\Omega\})$. The simplest example is a function of the form $f_1(u) = u$.

The specificity of the information granule Ω associated with its well-defined semantics (meaning) can be articulated in terms of the length of the interval. In the case of $\Omega = [a, b]$, any continuous nonincreasing function f_2 of the length of this interval, that is, $f_2(m(\Omega))$ where $m(\Omega) = |b-a|$ can serve as a sound indicator

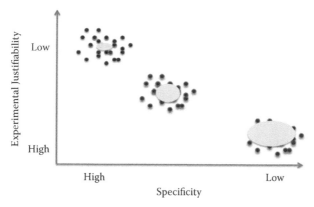

FIGURE 5.1
Specificity versus experimental evidence of interval information granules: the conflicting nature of the requirements. High experimental evidence: a lack of specificity. Low experimental evidence: high specificity. Relatively high specificity: acceptable experimental evidence.

of the specificity of the information granule. The shorter the interval (the higher the value of $f_2(m(\Omega))$), the better the satisfaction of the specificity requirement.

It is evident that the two requirements identified above are in conflict: the increase in the values of the criterion of experimental evidence (justifiability) comes at an expense of a deterioration of the specificity of the information granule (specificity). As illustrated in Figure 5.1, the increase in the evidence behind the information granule is associated with the decrease in its specificity. As usual, we are interested in forming a sound compromise between these requirements.

Having these two criteria in mind, let us proceed with the detailed formation of the interval information granule. We start with a numeric representative of the set of data **D** around which the information granule Ω is created. A sound numeric representative of the data is their median, med(**D**). Recall that the median is a robust estimator of the sample and typically comes as one of the elements of **D**. Once the median has been determined, Ω (the interval [a,b]) is formed by specifying its lower and upper bounds, denoted here by "a" and "b," respectively (refer also to Figure 5.2).

The determination of these bounds is realized independently. Let us concentrate on the optimization of the upper bound (b). The optimization of the lower bound (a) is carried out in an analogous fashion. For this part of the interval, the length of Ω or its nonincreasing function, as noted above is expressed as $|\text{med}(D) - b|$. In the calculations of the cardinality of the information granule, we take into consideration the elements of **D** positioned to the right from the median, that is, card$\{x_k \in D|$ med$(D) \le x_k \le b\}$. Again, in general, we can compute $f_1(\text{card}\{x_k \in D|$ med$(D) \le x_k \le b\})$, where f_1 is an increasing function. As the requirements of experimental evidence (*justifiable granularity*) and specificity (*semantics*) are in conflict, we can either resort to a certain

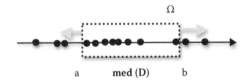

FIGURE 5.2
Optimization of interval information granule Ω by adjustment of its bounds.

version of multiobjective optimization and look at the resulting Pareto front or consider a maximization of the composite multiplicative index that is realized independently for the lower and upper bound of the interval, that is,

$$V(b) = f_1(\text{card}\{x_k \in D | \text{ med}(D) \leq x_k \leq b\}) \cdot f_2(|\text{med}(D) - b|) \qquad (5.1)$$

$$V(a) = f_1(\text{card}\{x_k \in D | a \leq x_k \leq \text{med}(D)\}) \cdot f_2(|\text{med}(D) - a|) \qquad (5.2)$$

We obtain the optimal upper bound b_{opt}, by maximizing the value of $V(b)$, namely $V(b_{opt}) = \max_{b > \text{med}(D)} V(b)$. In the same way, the lower bound a is constructed of the information granule, a_{opt}, that is, $V(a_{opt}) = \max_{a < \text{med}(D)} V(a)$.

Among many possible design options regarding functions f_1 and f_2, we consider the following alternatives:

$$f_1(u) = u \qquad (5.3)$$

$$f_2(u) = \exp(-\alpha u) \qquad (5.4)$$

where α is a positive parameter supporting some flexibility when optimizing the information granule Ω. Its essential role is to calibrate an impact of the specificity criterion on the constructed information granule. Note that if $\alpha = 0$ then $f_2(u) = 1$ and thus the criterion of specificity of information granule is completely ruled out (ignored). In this case, $b = x_{max}$ with x_{max} being the largest element in D. Higher values of α stress the increasing importance of the specificity criterion. Sufficiently high values of α promote very confined, numeric-like information granules.

As indicated, there is a substantial flexibility in defining f_1 and f_2. For instance a power function $f_1(u) = u^\beta$, $\beta > 0$ offers a higher level of flexibility in quantifying the impact of the first component on the values of the performance index (Equations 5.1 and 5.2).

Alluding to the performance indexes, it could be worth noting that the expression links to the concept of quantiles used in statistics but the construct investigated here becomes semantically richer. The first factor shown in this expression where $f_1(u) = u$ could be regarded as a mass of probability associated with the interval (as links to the property of the sufficient experimental evidence). The second factor underlines the need for the well-stressed semantics of the constructed information granule. In this way, we intend to capture two equally important features of the information granule being formed.

5.1.1 Some Illustrative Examples

In what follows, we present a series of numeric experiments, in which we use data that are drawn from some probability density functions as well as some finite collections of numeric data. We concentrate only on the optimization of the upper bound of the interval (b).

Data Governed by Some Underlying Probability Density Function. If the data are governed by some probability density function (pdf), then the optimization criterion takes this into consideration leading to the following expression: $V(b) = \int_0^b p(x)dx * \exp(-\alpha x)$. In the sequel, we obtain the optimal value of the bound by plotting this performance index V(b) regarded as a function of "b." An example of V(b) for several types of pdfs, namely uniform, Gaussian, and exponential ones is shown in Figure 5.3. We note that with the increase in the values of α, the specificity of the resulting information granule becomes higher (namely, the interval tends to be narrow). For instance, in the case of the Gaussian distribution, we see that $b_{opt} = 1.0$ and $b_{opt} = 0.5$ for the values of α equal to 0.6 and 2.0, respectively.

For comparison, we include the plots in which for the same pdfs we specify f_1 to be in the form $f_1(u) = u^2$ and $f_1(u) = u^{0.5}$ (see Figures 5.4 and 5.5). Note that the optimal values of "b" are shifted toward higher values for $f_1(u) = u^2$ and lower values for $f_1(u) = u^{0.5}$.

Another comprehensive view of the resulting construct can be obtained when plotting the values of f_1 versus f_2 (see Figure 5.6). It forms another presentation of the results contained in Figure 5.5. For the lower values of α, there is a section of the curve where f_2 changes leaving f_1 almost unchanged. The tendency changes with an increase in the values of α; a symmetrical relationship is observed for $\alpha = 0.4$ (see Figure 5.6c).

Finite Sets of Numeric Data. Now let us consider an example involving a finite number of one-dimensional data $D = \{3.1, 2.3, 1.7, 0.4, 1.9, 3.6, 4.0\}$. The median is equal to 2.3. The plots of the performance index are shown in Figure 5.7. Again a clearly delineated maximum of V(b) is present. The form of this dependency is associated with the discrete values of the elements in D hence a collection of decreasing segments of V. The corresponding intervals for selected values of α are the following:

$$\alpha = 0.0 \ [0.40 \ 4.00]$$

$$\alpha = 1.0 \ [1.69 \ 3.61]$$

$$\alpha = 1.5 \ [1.69 \ 2.32]$$

$$\alpha = 4.0 \ [1.88 \ 2.32]$$

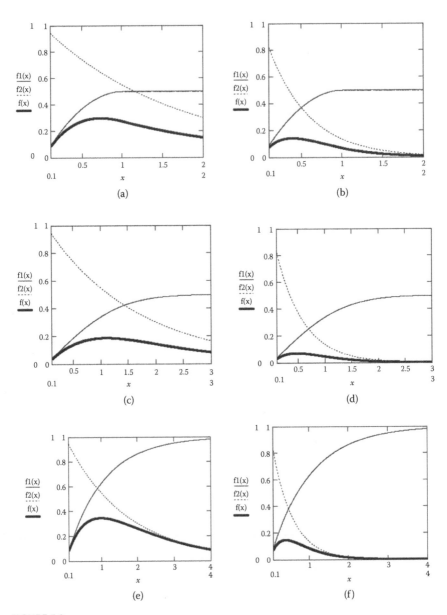

FIGURE 5.3

Plots of V(b) treated as a function of "b" exhibiting some maxima for the data governed by selected pdfs and selected values of α. (a) Uniform pdf over [0,1], α = 0.6. (b) Uniform pdf over [0,1], α = 2.0. (c) Gaussian pdf N(0, 2), α = 0.6. (d) Gaussian pdf N(0, 2), α = 2.0. (e) Exponential pdf, α = 0.6. (f) Exponential pdf, α = 2.0.

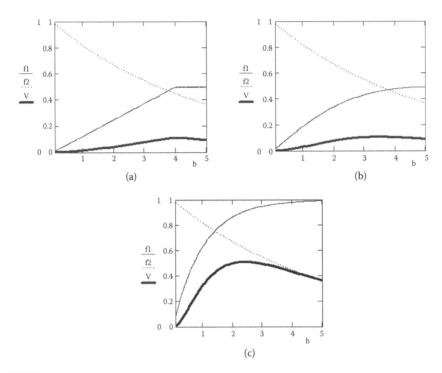

FIGURE 5.4
Plots of V(b) treated as a function of "b" for $f_1(u) = u^2$ for the data governed by selected pdfs. (a) Uniform pdf over [0,4]. (b) Gaussian pdf N(0, 2). (c) Exponential, $\alpha = 0.2$.

5.1.2 A Determination of Feasible Values of α

Alluding to the format of the maximized multiplicative objective function used in the optimization problem, it is insightful to elaborate on the choice of numeric values for the parameter of the construct that is the maximal value of α. Note that for $\alpha = 0.0$, we have $f_2(u)$ equal identically to 1 and only the first component of V(b) is used in the formation of the information granule; naturally the resulting interval includes all experimental data. Assume that the following ordering has been made $m < x_1 < x_2 < ....< x_n$. With regard to the possible range of α, a certain estimation process can be carried out. As noted earlier, $\alpha = 0$ entails that the coverage criterion is the only one involved and the interval becomes equal to $[x_{min}, x_{max}]$. The largest value of α, that is, α_{max}, is determined by requesting that the optimal interval is as narrow as possible. In the case of the upper bound, we request that the interval is $[m, x_1]$ where x_1 is the closest to "m" and $x_1 > m$. This happens for such value of α, call it α_{max}, for which the following set of inequalities are satisfied:

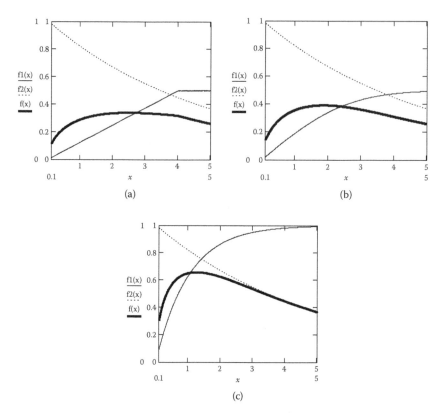

FIGURE 5.5
Plots of V(b) treated as a function of "b" for $f_1(u) = u^{0.5}$ for the data governed by selected pdfs. (a) Uniform pdf over [0,4]. (b) Gaussian pdf N(0, 2). (c) Exponential, $\alpha = 0.2$.

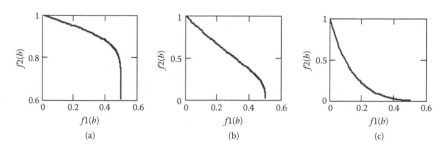

FIGURE 5.6
Plots of relationships f_1-f_2 (Gaussian pdf N(0, 2)). (a) $\alpha = 0.05$. (b) $\alpha = 0.4$. (c) $\alpha = 1.5$.

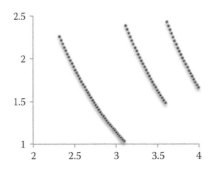

FIGURE 5.7
Plot of V(b) used in the determination of the upper bounds of the interval; $\alpha = 1$.

$$1^{*}\exp(-\alpha|m\text{-}x_1|) > 2^{*}\exp(-\alpha|m-x_2|)$$

$$1^{*}\exp(-\alpha|m\text{-}x_1|) > 3^{*}\exp(-\alpha|m-x_3|)$$

$$\ldots$$

$$1^{*}\exp(-\alpha|m\text{-}x_1|) > l^{*}\exp(-\alpha|m-x_l|) \qquad (5.5)$$

We can look at these inequalities one by one and solve them for the maximal values of α. By taking the first inequality, we have $\exp(\alpha(|m\text{-}x_1|-|m\text{-}x_2|)) = 1/2$ and $\alpha_{max}(1) = \dfrac{\ln(1/2)}{|m-x_1|-|m-x_2|}$. By solving successive equations, we obtain $\alpha_{max}(2) = \dfrac{\ln(1/3)}{|m-x_1|-|m-x_2|} \ldots \alpha_{max}(l\text{-}1) = \dfrac{\ln(1/l)}{|m-x_1|-|m-x_2|}$ and finally $\alpha_{max} = \max(\alpha_{max}(1), \alpha_{max}(2)\ldots \alpha_{max}(l\text{-}1))$.

As a result we obtain the range $[0, \alpha_{max}]$. Note that for the determination of α_{max} for x_i's lower than "m" and contributing to the formation of the lower bound (a), the corresponding maximal value of α may result in a different maximal value of α.

To unify these ranges of the values of the parameters determined for individual cases, we normalize each of them to [0,1] in the form α/α_{max} so that when specifying a single value of α, we construct the corresponding bounds of the interval. This construct is helpful when forming intervals with the use of a single external parameter.

The construction of the lower bound of Ω is realized in the same manner as discussed above. There are two different values of α_{max} associated with the lower and upper bound, call them $\alpha_{max}(a)$ and $\alpha_{max}(b)$, however, from the outside, one uses a single parameter α with the values confined to the unit

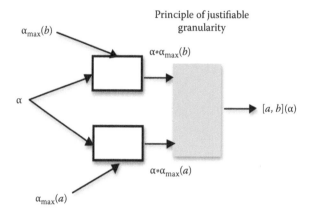

FIGURE 5.8
Building an interval information granule with the use of a single parameter α through the internal normalization mechanism.

interval. When used internally in the construction, the interval used is a renormalized value equal to $\alpha*\alpha_{max}(a)$ and $\alpha*\alpha_{max}(b)$ (see Figure 5.8).

5.1.3 Formation of Shadowed Sets or Rough Sets of Information Granules

Let us note that in spite of their fundamental differences, the concepts of shadowed sets (Pedrycz, 2009) or rough sets (Pawlak, 1982) share some conceptual similarity in the sense that these information granules identify three regions in the universe of discourse: (a) full membership (complete belongingness), (b) full exclusion (lack of belongingness), and (c) ignorance about membership (no knowledge about membership of elements located in this region is available). The plot presented in Figure 5.9 illustrates the essence of the construct.

The objective function takes into account the different nature of the regions by bringing them into the overall expression using a certain weighting scheme, which looks differently at the region of full membership and ignorance. More specifically, we admit that the region characterizing a lack

$a - da \qquad a \qquad\qquad\qquad b \quad b + db$

FIGURE 5.9
An example of a shadowed set or rough set with regions of full membership, exclusion, and ignorance about the concept of membership.

of membership knowledge should contribute to a different extent when counting elements falling within this region (the count has to be discounted). Likewise we discount the length of the information granule when dealing with this region. Considering the optimization of the bounds "b" and "db" (and effectively the sum b+db), we have

$$V(b, db) = f_1(card\{x_k \in D | \ med(D) \leq x_k \leq b\} + \gamma card\{x_k \in D | \ b < x_k \leq b+db\}).$$
$$.f_2(|med(D)-b|+\gamma|b+db-b|) \tag{5.6}$$

where γ denotes a weight coefficient assuming values lower than 1 (and thus discounting the regions of the shadow); in particular, we can set its value to ½. The maximization of Equation (5.6) gives rise to the optimal values of "b" and "c"(= b+db) or alternatively "b" and "db."

As an example, we consider data governed by the uniform pdf over [–4, 4] for which we form a shadowed set around the numeric representative equal to zero (see Figure 5.10). The intent is to maximize V(b,db) with "b" and "db" being the parameters of the shadowed set. As before we optimize only the upper bound of the shadowed set with its two parameters. The plots of V treated as a function of "b" and "db" for selected values of $\gamma = 0.2$ and 0.5 are shown in Figure 5.10 with the optimal values of b = 0.368, db = 0.66 and b = 0.205, db = 0.59.

5.1.4 Weighted Data in the Construction of Information Granules

The above construct of an information granule can be generalized to cope with situations where the individual data are associated with some weights (which could quantify their quality which may vary from one element to another). Given data in the form (x_i, w_i) where the weights w_i assume values

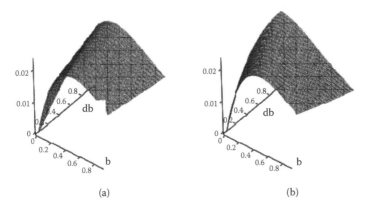

(a)   (b)

FIGURE 5.10
Optimized performance index V(b, db) as a function of "b" and "db" for $\gamma = 0.2$ (a) and 0.5 (b).

located in the [0,1] interval, $\mathbf{w} = [w_1, w_2, \ldots, w_N]$, we reformulate the maximized performance index to be in the form

$$V(b) = f_1\left(\sum_{\substack{k=1 \\ x_k:\, \text{med} < x_k < b}}^{N} w_k \right) f_2(|b - \text{med}(\mathbf{D}, \mathbf{w})|) \qquad (5.7)$$

where med($\mathbf{D}, \mathbf{w}$) is a weighted median (whose computing uses the weighted data). Recall that the weighted median is found as a solution to the minimization problem

$$\text{med}(\mathbf{D}, \mathbf{w}) = \text{argmin}_y \sum_{k=1}^{N} w_k |x_k - y| \qquad (5.8)$$

5.1.5 From a Family of Interval Information Granules to a Fuzzy Set

The parameter α used in the construction of the interval information granule plays a pivotal role in the formation of a fuzzy set as a justifiable information granule. Likewise we can directly use the bounds of this parameter in the formation of the fuzzy set.

Following the discussion concerning the normalization of the values of α to the unit interval and in virtue of the monotonicity of the interval granules with regard to the values of α, each information granule can be associated with the corresponding value of α in [0,1], that is, $\Omega(\alpha)$. For different values of α, a collection of the corresponding $\Omega(\alpha)$s form a nested family of intervals that can be regarded as a family of α-cuts of a certain fuzzy set Ω with the membership function resulting from the representation theorem.

5.1.6 Development of Fuzzy Sets of Type 2

Fuzzy sets are described by *numeric* membership functions. In contrast, probabilistic sets, type-2 fuzzy sets, interval-valued fuzzy sets, or *granular fuzzy sets*, in general, generalize fuzzy sets. There is a significant conceptual departure: granular fuzzy sets with membership grades modeled as information granules are more in rapport with reality by departing from the sometimes too demanding requirement of reliance on numeric membership grades. We can witness many studies devoted to type-2 fuzzy sets (which are one of the visible realizations of granular fuzzy sets). There is a wealth of their applications. Interestingly enough, these studies do not raise and solve a central problem of forming granular values of membership, which brings some difficulty at the application end of the spectrum. Let us note that having a collection of fuzzy sets of type 1 (namely, with numeric membership functions) defined in the same universe of discourse, by applying the principle of justifiable granularity we can easily determine

TABLE 5.1

A Collection of Fuzzy Sets and the Resulting Type-2 (Interval-Valued) Fuzzy Set

	x_1	x_2	x_3	x_4
A_1	0.10	0.40	0.70	1.00
A_2	0.30	0.30	0.80	0.90
A_3	0.21	0.28	0.85	0.80
A_4	0.25	0.42	0.71	0.85
A_5	0.05	0.34	0.65	0.78
Type-2 fuzzy set	[0.10 0.25]	[0.30 0.40]	[0.70 0.80]	[0.80 0.90]

the granular membership grades, that is, intervals, thus forming an interval-valued fuzzy set.

As an example, let us consider a collection of fuzzy sets A_1, A_2,..., A_5 defined in the finite universe of discourse (see Table 5.1). By adopting the median as a numeric representative for each collection of membership values (for each element of the universe of discourse) and maximizing the performance index Equations (5.1)–(5.2) with $\alpha = 1.0$; the resulting bounds are presented in Table 5.1.

5.1.7 A Design of Multidimensional Information Granules

The realized information granule (either in the form of an interval or a fuzzy set) exploits one-dimensional numeric data. The extension to the multidimensional case is straightforward and can be done by constructing a Cartesian product of the information granules formed for the individual variables (i.e., the construct is realized for the individual variables of the multidimensional data). For instance, given an information granule A defined in X_1, B arising at X_2, and C at X_3, the result is in the form A × B × C. In the case of the interval information granules, we take the minimum of the successive characteristic functions $(A \times B \times C)(x_1, x_2, x_3) = \min(A(x_1), B(x_2), C(x_3))$. Examples of Cartesian products—information granules forming *boxes* over the two-dimensional universe of discourse and viewed as *granular* prototypes are illustrated in Figure 5.11.

5.1.8 A General View of the Principle of Information Granularity: A Diversity of Formal Setups of Information Granularity

It is worth stressing that the principle of justifiable granularity covers a broad spectrum of scenarios—all of them can be arranged along the two coordinates. The first one is concerned with the formal environment of information granules. The second one points at the nature of available experimental evidence used in the realization of the information granule (sets, fuzzy sets, etc.).

In our considerations so far we have focused on the use of numeric evidence while constructing interval-like information granules and those

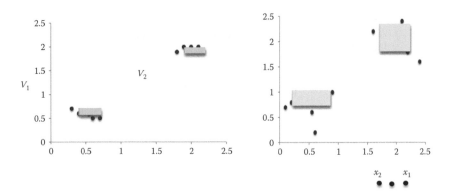

FIGURE 5.11
Granular prototypes of the data.

expressed by fuzzy sets. The principle developed covers other cases; some modifications of the criteria pertinent to the specificity of the contemplated realization are to be envisioned. Information granules can be formed for the data which themselves are granular, that is, intervals, fuzzy sets, and so forth. The main flow of the construction is the same as before, however, the objective function needs to be carefully revisited to reflect the essence of the information granules one has to deal with. Let us consider intervals, fuzzy sets, and probability density functions (see Figure 5.12) and look for the information granule described in terms of a certain interval.

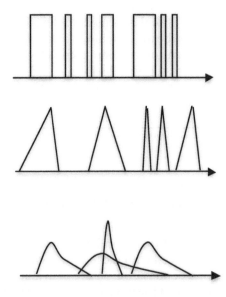

FIGURE 5.12
Examples of granular data along with the information granule: intervals, fuzzy sets, and probability density functions.

The second component (quantifying the specificity of the information granule) of the performance index (Equations 5.1 and 5.2) remains the same. The first component of the objective function V is modified as follows (in the following formulas we are concerned with the upper bound of the interval),

for intervals,

$$V(b) = f_1 \left(\sum_{i=1}^{N} \int_X X(x) X_i(x) dx \right) \tag{5.9}$$

for fuzzy sets,

$$V(b) = f_1 \left(\sum_{i=1}^{N} Poss(X, X_i) \right) \tag{5.10}$$

where $Poss(X, X_i)$ denotes a possibility measure (Dubois and Prade, 1980) of X with regard to X_i, which expresses the degree of overlap of these two information granules. Proceeding with the detailed calculations, we have Poss (X, X_i) = $sup_x [min(X(x), X_i(x))]$,

for probability density functions,

$$V(b) = f_1 \left(\sum_{i=1}^{N} \int_X X(x) p_i(x) dx \right) \tag{5.11}$$

where $p_i(x)$ is a pdf of the *i*-th probabilistic information granule and the integral is the expected value of X taken with respect to the *i*-th probabilistic information granule described by its pdf (which results in the probability of the event X).

The principle of justifiable granularity covers a broad spectrum of formal setups of information granules—all of them can be arranged along two coordinates, Figure 5.13. The first one is concerned with the formal environment of information granules. The second one points at the nature of the experimental evidence (sets, fuzzy sets, etc.) used with which the information granules are constructed.

5.2 Construction of Information Granules through Clustering of Numeric Experimental Evidence

Clustering delivers a natural mechanism to construct information granules in the presence of numeric data. As a matter of fact, a main agenda of clustering is to reveal a structure of data, namely, form a collection of

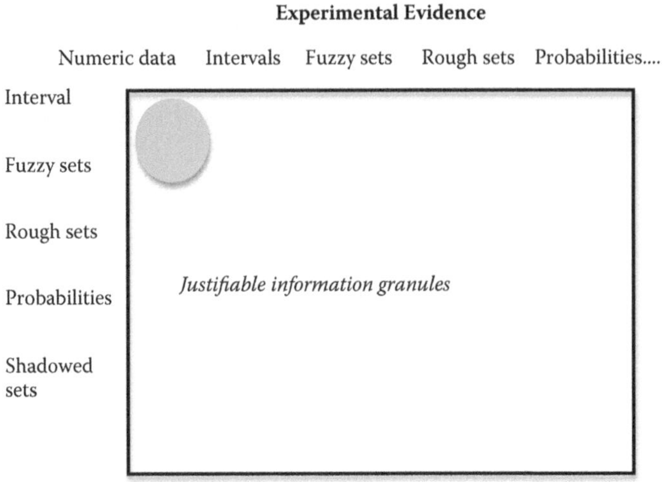

Experimental Evidence

Numeric data   Intervals   Fuzzy sets   Rough sets   Probabilities....

Interval

Fuzzy sets

Rough sets

Probabilities   *Justifiable information granules*

Shadowed
sets

FIGURE 5.13

The principle of justifiable granularity: a variety of environments of information granules and the nature of experimental evidence on the basis of which information granules are designed.

clusters—information granules. There is a genuine diversity of clustering algorithms. Depending upon the method used, the results arise as information granules expressed in terms of sets, fuzzy sets, rough sets, and so forth. In what follows, we discuss one of the commonly used algorithms of clustering which produces a collection of fuzzy sets, namely, fuzzy c-means (FCM) (Bezdek, 1981). The method is well established with a wealth of applications and algorithmic enhancements. Furthermore, it forms a sound generalization of the K-means clustering technique that forms set-based information granules.

Let us briefly review the formulation of the FCM, develop the algorithm, and highlight the main properties of the fuzzy clusters. Given a collection of n-dimensional data sets $\{x_k\}$, k = 1,2,...,N, the task of determining its structure—a collection of "c" clusters, is expressed as a minimization of the following objective function (performance index) Q being regarded as a sum of the squared distances

$$Q = \sum_{i=1}^{c} \sum_{k=1}^{N} u_{ik}^{m} \| x_k - v_i \|^2 \qquad (5.12)$$

where $v_1, v_2, ..., v_c$ are n-dimensional prototypes of the clusters and $U = [u_{ik}]$ stands for a partition matrix expressing a way of allocation of the data to the corresponding clusters; u_{ik} is the membership degree of data x_k in the i-th cluster. The distance between the data x_k and prototype v_i is denoted by $\|.\|$.

The fuzzification coefficient m (>1.0) expresses the impact of the membership grades on the individual clusters and produces certain geometry of the information granules.

A partition matrix satisfies two important properties,

$$
\text{(a)} \quad 0 < \sum_{k=1}^{N} u_{ik} < N, \quad i = 1, 2, ..., c \tag{5.13}
$$

$$
\text{(b)} \quad \sum_{i=1}^{c} u_{ik} = 1, \quad k = 1, 2, ..., N \tag{5.14}
$$

Let us denote by **U** a family of matrices satisfying these two requirements (a) and (b). The first requirement states that each cluster has to be nonempty and different from the entire set. The second requirement states that the sum of the membership grades should be confined to 1.

The minimization of Q completed with respect to $U \in \mathbf{U}$ and the prototypes $\mathbf{v}_i$ of $V = \{\mathbf{v}_1, \mathbf{v}_2, ... \mathbf{v}_c\}$ of the clusters. More explicitly, we write it down as follows:

$$
\min Q \text{ with respect to } U \in \mathbf{U}, \mathbf{v}_1, \mathbf{v}_2, ..., \mathbf{v}_c \in \mathbf{R}^n \tag{5.15}
$$

From the optimization standpoint, there are two individual optimization tasks to be carried out separately for the partition matrix and the prototypes. The first one concerns the minimization with respect to the constraints given the requirement of the form (Equations 5.13 and 5.14), which holds for each data point x_k. The use of Lagrange multipliers transforms the problem into its constraint-free version. The augmented objective function formulated for each data point, k = 1, 2, ..., N, reads as

$$
V = \sum_{i=1}^{c} u_{ik}^m d_{ik}^2 + \lambda \left(\sum_{i=1}^{c} u_{ik} - 1 \right) \tag{5.16}
$$

where we use an abbreviated form $d_{ik}^2 = ||\mathbf{x}_k - \mathbf{v}_i||^2$. Proceeding with the necessary conditions for the minimum of V for k = 1, 2, ...N, one has

$$
\frac{\partial V}{\partial u_{st}} = 0 \quad \frac{\partial V}{\partial \lambda} = 0 \tag{5.17}
$$

s = 1, 2... c, t = 1, 2...N. Now we calculate the derivative of V with respect to the elements of the partition matrix in the following way:

$$
\frac{\partial V}{\partial u_{st}} = mu_{st}^{m-1} d_{st}^2 + \lambda \tag{5.18}
$$

From Equation (5.17) and the use of the normalization condition (Equation 5.14), we calculate the membership grade u_{st} to be equal to

$$u_{st} = -\left(\frac{\lambda}{m}\right)^{\frac{1}{m-1}} d_{st}^{\frac{2}{m-1}} \tag{5.19}$$

We complete some rearrangements of the above expression by isolating the term including the Lagrange multiplier

$$-\left(\frac{\lambda}{m}\right)^{\frac{1}{m-1}} = \frac{1}{\sum_{j=1}^{c} d_{jt}^{\frac{2}{m-1}}} \tag{5.20}$$

Inserting this expression into Equation (5.19), we obtain the successive entries of the partition matrix

$$u_{st} = \frac{1}{\sum_{j=1}^{c} \left(\frac{d_{st}^2}{d_{jt}^2}\right)^{\frac{1}{m-1}}} \tag{5.21}$$

The optimization of the prototypes v_i is carried out assuming the Euclidean distance between the data and the prototypes, that is, $||x_k - v_i||^2 = \sum_{j=1}^{n} (x_{kj} - v_{ij})^2$. The objective function reads now as follows: $Q = \sum_{i=1}^{c} \sum_{k=1}^{N} u_{ik}^m \sum_{j=1}^{n} (x_{kj} - v_{ij})^2$ and its gradient with respect to v_i, $\nabla_{v_i} Q$ made equal to zero yields the system of linear equations

$$\sum_{k=1}^{N} u_{ik}^m (x_{kt} - v_{st}) = 0, s = 1, 2,.., c\ t = 1,2,..., n \tag{5.22}$$

Thus,

$$v_{st} = \frac{\sum_{k=1}^{N} u_{ik}^m x_{kt}}{\sum_{k=1}^{N} u_{ik}^m} \tag{5.23}$$

One should emphasize that the use of some other distance function different from the Euclidean one brings some computational complexity and the formula for the prototype cannot be presented in the concise manner as given above.

Overall, the FCM clustering is completed through a sequence of iterations where we start from some random allocation of data (a certain randomly initialized partition matrix) and carry out the following updates by successively adjusting the values of the partition matrix and the prototypes. The iterative process is repeated until a certain termination criterion has been satisfied. Typically, the termination condition is quantified by looking at the changes in the membership values of the successive partition matrices. Denote by U(t) and U(t+1) the two partition matrices produced in two consecutive iterations of the algorithm. If the distance $||U(t+1)-U(t)||$ is less than a small predefined threshold ε, then we terminate the algorithm. Typically, one considers the Tchebyschev distance between the partition matrices meaning that the termination criterion reads as follows:

$$\max_{i,k} |u_{ik}(t+1) - u_{ik}(t)| \le \varepsilon \qquad (5.24)$$

The fuzzification coefficient exhibits a direct impact on the geometry of fuzzy sets generated by the algorithm. Typically, the value of "m" is assumed to be equal to 2.0. Lower values of m (that are closer to 1) yield membership functions that start resembling characteristic functions of sets; most of the membership values become localized around 1 or 0. The increase in the fuzzification coefficient (m = 3, 4, etc.) produces *spiky* membership functions with the membership grades equal to 1 at the prototypes and a fast decline of the values when moving away from the prototypes. Several illustrative examples of the membership functions are included in Figure 5.14. Here the prototypes are equal to 1, 3.5, and 5 while the fuzzification coefficient assumes values of 1.2 (a), 2.0 (b), and 3.5 (c). In addition to the varying shape of the membership functions, observe that the requirement put on the sum of membership grades imposed on the fuzzy sets yields some rippling effect: the membership functions are not unimodal but may exhibit some ripples whose intensity depends upon the distribution of the prototypes and the values of the fuzzification coefficient. The intensity of the rippling effect is also affected by the values of "m" and increases with the higher values of "m."

In case m tends to 1, the partition matrix comes with the entries equal to 0 or 1 so the results come in the form of the Boolean partition matrix. The algorithm becomes then the well-known K-means clustering.

From the notation perspective, let us note that the partition matrix U can be sought as a collection of membership functions of the information granules occupying successive rows of the matrix. In other words, fuzzy sets A_1, A_2,.., A_c with their membership values assumed for the data x_1, x_2, ..., x_N form the corresponding rows of U, that is,

$$U = \begin{bmatrix} A_1 \\ A_2 \\ ... \\ A_c \end{bmatrix} = \begin{bmatrix} \mathbf{u}_1 \\ \mathbf{u}_2 \\ ... \\ \mathbf{u}_c \end{bmatrix}$$

where $A_i(x_k) = u_{ik}$ and $\mathbf{u}_i$ stands for a vector of membership degrees in cluster "i."

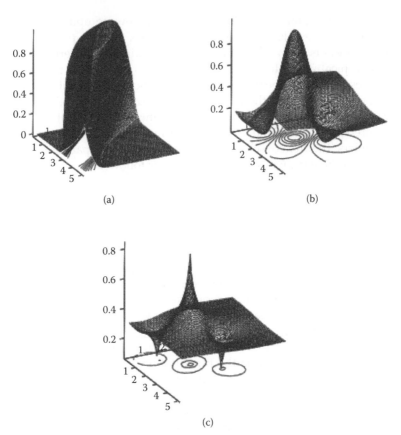

FIGURE 5.14
Examples of membership functions of fuzzy sets in R2 for selected values of "m" (a) m = 1.2, (b) m = 2.0, and (c) m = 3.0.

There is a genuine plethora of generalizations of the FCM method, both in terms of its conceptual extensions or algorithmic (optimization) enhancements. From the perspective of Granular Computing, an interesting option concerns clustering of objects that are information granules themselves rather than vectors of real numbers. In this case, information granules need to be represented in a certain space and their descriptors are clustered afterward. Evidently, the results of clustering, especially prototypes (representatives) are information granules as well.

As discussed in the literature (Hathaway, Bezdek, and Pedrycz, 1996; Pedrycz et al., 1998), with this regard the two approaches are encountered:

(a) *Parametric*. Here it is assumed that information granules to be clustered are represented in a certain parametric format, that is, they are described by triangular fuzzy sets, intervals, Gaussian membership functions, and so forth. In each case there are several parameters associated with the membership functions and the clustering is carried out in the space of the parameters. The resulting prototypes are also described by information granules having the same parametric form as the information granules being clustered. Obviously, the dimensionality of the space of the parameters in which the clustering takes place is usually higher than the original space. For instance, in the case of triangular membership functions, the new space is $\mathbf{R}^{3n}$ (given the original space is $\mathbf{R}^n$).

(b) *Nonparametric*. There is no particular form of the information granules so some characteristic descriptors of information granules used to capture the nature of these granules are formed and used to carry out clustering.

Information granules A_i are multivariable in the sense they are formed over $\mathbf{R}^n$. In light of this multidimensional nature of these constructs, their interpretability could be limited. To alleviate this difficulty, one can build a collection of one-dimensional information granules by projecting the prototypes on the individual coordinates (variables). For the j-th variable we obtain the prototypes $v_{1j}, v_{2j}, \ldots, v_{cj}$. As real numbers, they can be ordered and assigned some semantics. If those are ordered in the increasing order, then we can assign some labels (information granules) to them which convey a clear semantics, that is, negative *large*, negative *medium*, negative *small*,…, positive *large*. Proceeding with the same projection for the remaining variables, we translate the prototypes into a Cartesian product of the one-dimensional information granules. An example of this construction is presented in Figure 5.15. Here we have four information granules constructed in the two-dimensional space and the projection results into the information granules in the form: (NS, NS) (PL, PS) (PM, PM) (PM, PL) where the information granules have been labeled as *Negative Small* (NS), *Positive Small* (PS), *Positive Medium* (PM), and *Positive Large* (PL) for x_1 and for x_2. For the two information granules, $\mathbf{v}_2$ and $\mathbf{v}_3$, for one of the coordinates the descriptors are the same (PM).

5.3 Knowledge-Based Clustering: Bringing Together Data and Knowledge

There is some domain knowledge and it has to be carefully incorporated into the generic clustering procedure. Knowledge hints can be conveniently captured and formalized in terms of fuzzy sets. Altogether with the underlying clustering algorithms, they give rise to the concept of knowledge-based clustering—a unified framework in which data and knowledge are processed together in a uniform fashion.

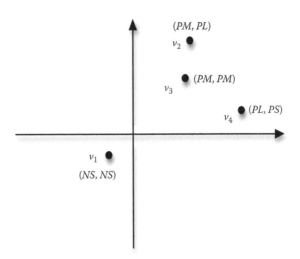

FIGURE 5.15
Four information granules and their projection along with the construction of the Cartesian product of the one-dimensional information granules.

We discuss some of the typical design scenarios of knowledge-based clustering and show how the domain knowledge can be effectively incorporated into the fabric of the original data-driven only clustering techniques.

We can distinguish several interesting and practically viable ways in which domain knowledge is taken into consideration:

A subset of labeled patterns. The knowledge hints are provided in the form of a small subset of labeled patterns $K \subset N$ (Pedrycz and Waletzky, 1997). For each of them we have a vector of membership grades f_k, $k \in K$, which consists of degrees of membership the pattern is assigned to the corresponding clusters. As usual, we have $f_{ik} \in [0, 1]$ and $\sum_{i=1}^{c} f_{ik} = 1$.

Proximity-based clustering. Here we are provided a collection of pairs of patterns (Loia, Pedrycz, and Senatore, 2007) with specified levels of closeness (resemblance), which are quantified in terms of proximity, prox (k, l) expressed for x_k and x_l. The proximity offers a very general quantification scheme of resemblance: we require reflexivity and symmetry, that is, prox(k, k) = 1 and prox(k, l) = prox(l, k), however, no transitivity is needed.

"Belong" and "not-belong" Boolean relationships between patterns. These two Boolean relationships stress that two patterns should belong to the same clusters, $R(x_k, x_l) = 1$ or they should be placed apart in two different clusters, $R(x_k, x_l) = 0$. These two requirements could be relaxed by requiring that these two relationships return values close to one or zero.

Uncertainty of labeling/allocation of patterns. We may consider that some patterns are *easy* to assign to clusters while some others are inherently difficult to deal with meaning that their cluster allocation is associated with a significant level of uncertainty. Let $F(x_k)$ stand for the uncertainty measure (e.g., entropy) for x_k (as a matter of fact, F is computed for the membership degrees of x_k, that is, $F(u_k)$ with u_k being the k-th column of the partition matrix). The uncertainty hint is quantified by values close to 0 or 1 depending upon what uncertainty level a given pattern is coming from.

Conditional (context-based) clustering. In this scenario, a construction of information granules is completed in a given context defined in a certain auxiliary variable while the context itself is an information granule. We can think of this clustering as a more focused, oriented realization of information granules. Consider that over the auxiliary variable provided is a context (fuzzy set) B. The essence of the construct is to impose a context on the partition matrix so that the information granules are directly implied by it.

Depending on the character of the knowledge hints, the original clustering algorithm needs to be properly refined. In particular, the underlying objective function has to be augmented to capture the knowledge-based requirements. Shown below are several examples of the extended objective functions dealing with the knowledge hints introduced above.

When dealing with some labeled patterns, we consider the following augmented objective function

$$Q = \sum_{i=1}^{c} \sum_{k=1}^{N} u_{ik}^m \| x_k - v_i \|^2 + \beta \sum_{i=1}^{c} \sum_{k=1}^{N} (u_{ik} - f_{ik} b_k)^2 \| x_k - v_i \|^2 \qquad (5.25)$$

where the second term quantifies distances between the class membership of the labeled patterns and the values of the partition matrix. The positive weight factor (b) helps set up a suitable balance between the knowledge about classes already available and the structure revealed by the clustering algorithm. The Boolean variable b_k assumes values equal to 1 when the corresponding pattern has been labeled.

The proximity constraints are accommodated as a part of the optimization problem where we minimize the distances between proximity values being provided and those generated by the partition matrix $P(k_1, k_2)$

$$Q = \sum_{i=1}^{c} \sum_{k=1}^{N} u_{ik}^m \| x_k - v_i \|^2$$

$$\| prox(k_1, k_2) - P(k_1, k_2) \| \to Min \ k_1, k_2 \in K \qquad (5.26)$$

with K being a pair of patterns for which the proximity level has been provided. It can be shown that given the partition matrix, the expression $\sum_{i=1}^{c} min(u_{ik1}, u_{ik2})$ generates the corresponding proximity value.

For the uncertainty constraints, the minimization problem can be expressed as follows:

$$Q = \sum_{i=1}^{c} \sum_{k=1}^{N} u_{ik}^{m} \| \mathbf{x}_k - \mathbf{v}_i \|^2$$

$$\| F(u_k) - g_k \| \to \text{Min k } K \tag{5.27}$$

where K stands for the set of patterns for which we are provided with the uncertainty values g_k.

Proceeding with the context-based clustering, the form of the optimized objective function is the same as for the generic FCM, but the constraint imposed on the partition matrix comes with the constraint of the context. If the context B associated with the k-th data assumes the value b_k, the sum of the membership grades is equal to b_k instead of 1, namely $\sum_{i=1}^{c} u_{ik} = b_k$. The optimization scheme is completed in the same manner as for the FCM. The major difference in terms of the results shows up in the formula for the partition matrix U whose entries are computed as follows:

$$u_{ik} = \frac{b_k}{\sum_{j=1}^{c} \left(\dfrac{\| \mathbf{x}_k - \mathbf{v}_i \|}{\| \mathbf{x}_k - \mathbf{v}_j \|} \right)^{2/(m-1)}} \tag{5.28}$$

The calculations of the prototypes are carried out in the same way as in the generic version of the FCM.

Undoubtedly, the extended objective functions call for the optimization scheme that is more demanding as far as the calculations are concerned. In several cases, we cannot modify the standard technique of Lagrange multipliers, which leads to an iterative scheme of successive updates of the partition matrix and the prototypes. In general, though, the knowledge hints give rise to a more complex objective function in which the iterative scheme cannot be useful in the determination of the partition matrix and the prototypes. Alluding to the generic FCM scheme, we observe that the calculations of the prototypes in the iterative loop are doable in the case of the Euclidean distance. Even the Hamming or Tchebyshev distance brings a great deal of complexity. Likewise, the knowledge hints lead to the increased complexity: the prototypes cannot be computed in a straightforward way and one has to resort to more advanced optimization techniques. Evolutionary computing arises here as an appealing alternative. We may consider any of the options available there, including genetic algorithms (GAs), particle swarm optimization (PSO), ant colonies, to name some of them. The general scheme can be schematically structured as follows:

- repeat { EC (prototypes); compute partition matrix U;}

Note that the genetic optimization is concerned with the prototypes while the partition matrix is computed on their basis; by making this choice of the optimized parameters, we effectively cope with the search space of the lower dimensionality.

5.4 Refinement of Information Granules through Successive Clustering

The underlying idea is to successively split information granules, which are characterized by the most diversified (heterogeneous) content. Visually, we can portray the process of refining an information granule coming with the highest information diversity (information content) as illustrated in Figure 5.16.

Let us recall that when considering fuzzy clustering, that is, FCM (Pedrycz, 2005; Pedrycz and Gomide, 1998), the membership function of the information granule A_i at the highest level (ii = 1) is expressed as follows:

$$A_i[1](x) = \frac{1}{\sum_{j=1}^{c} \left(\frac{\| x - v_i[1] \|}{\| x - v_j[1] \|} \right)^{2/(m-1)}}$$

(5.29)

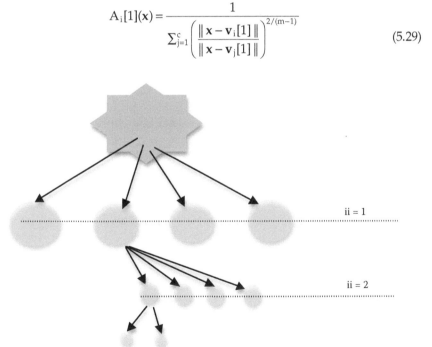

FIGURE 5.16
Refinement of information granule. The granule with the highest information content is split (refined) into more detailed information granules. The process is repeated and a hierarchy of information granules is constructed.

The notion of information content expressing the diversity of data falling under the realm of a given information granule plays a vital role in the realization of the hierarchical formation of the granules. Here we look at two alternatives on how the information content can be formulated.

Let us start with associating with the *i*-th cluster all data that belong to it to the highest extent (higher than to the remaining clusters) $X_i = \{x_k | \, u_{ik} = \max_j u_{jk}[1]\}$. The two alternatives under investigation are as follows:

Class membership content. This information content is of interest when the elements in the input space are associated with some classes, namely, each x_k belongs to one of the classes $\Omega_1, \Omega_2,.., \Omega_p$. The information content of this nature is of interest in the case of classification problems. We determine the class with the maximal number of data contained in X_i. Then we determine a classification error of this cluster by counting all data that do not belonging to this dominant class,

$$V_i = \text{classification error associated with the } i\text{-th cluster}$$

Output variable content. Considering that x_k is associated with some output y_k (so we are concerned with regression types of problems), we evaluate the content of the cluster from this perspective. We compute the numeric representative y^* of y_ks for all x_k which belong to X_i,

$$y^* = \frac{\sum_{x_k \in X_i} u_{ik}^m[1] y_k}{\sum_{x_k \in X_i} u_{ik}^m[1]} \tag{5.30}$$

In the sequel, the output content of this cluster is V_i expressed as a sum

$$V_i = \sum_{k:x_k \in X_k} (y_k - y^*)^2 \tag{5.31}$$

Any of the information content measures presented above can be used to guide the refinement of the clusters. The cluster "i_0" which characterizes by the highest information content $V_{i0} = \max_{i=1,2,...,c} V_i$ is subject to further refinement (splitting). The sum of V_is over all information granules describes an overall content of the information granules developed so far.

Considering that $A_i[ii]$ has been identified as having the highest value of information content V_i we split it into several successive information granules, that is, $A_1[ii+1], A_2[ii+1],... A_c[ii+1]$, so the structure of data refined by those information granules is formed as a result of the clustering algorithm. Given that the clustering has to be completed in the context of the information granule $A_i[ii]$ meaning that the requirement is expressed in the form $\sum_{i=1}^c A_i[ii+1](x) = A_i[ii](x)$, we note that this is nothing but an example of

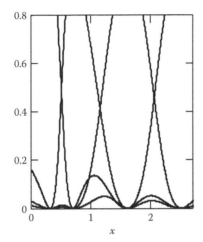

FIGURE 5.17
Plots of four membership functions.

context-based clustering. The membership degrees at the next level of specialization of information granule are expressed as

$$A_1[ii+1](x) = \frac{A_i[ii](x)}{\sum_{j=1}^{c} \left(\frac{\|x-v_1[ii]\|}{\|x-v_j[ii]\|}\right)^{2/(m-1)}}$$ (5.32)

As an illustrative example, consider a one-dimensional case in which we construct four fuzzy sets with the prototypes assuming the values $v_1[1] = 0.31$, $v_2[1] = 0.69$, $v_3[1] = 1.6$, $v_4[1] = 2.5$. The fuzzification coefficient is set to $m = 2.0$. The resulting membership functions are shown in Figure 5.17.

Assume that the third cluster is characterized by the highest value of the performance index and as such is further refined by forming three fuzzy sets centered around the prototypes $v_1[2] = 1.2$, $v_2[2] = 1.9$, $v_3[2] = 2.4$. Here the fuzzification coefficient used is set to 1.5. In virtue of the expression (Equation 5.32), the fuzzy sets formed there are subnormal with the membership functions shown in Figure 5.18.

5.5 Collaborative Clustering and Higher-Level Information Granules

In building information granules in scenarios where a number of individual data sets are processed individually, the data sets $\mathbf{D}_1$, $\mathbf{D}_2$, ..., $\mathbf{D}_p$ are defined in the same feature space. In many cases some collaborative pursuits are worth pursuing (see Pedrycz and Rai, 2009).

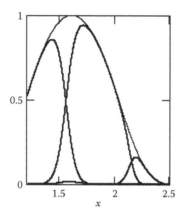

FIGURE 5.18
Plots of two fuzzy sets as a refinement of fuzzy set $A_3[1]$.

Two scenarios are envisioned. In the passive one, a global view of the structure of information granules is formed on the basis of the locally available data. In the active approach, the results locally obtained are further modified in light of the global assessment of the consistency between the local and global structures. In all detailed calculations, we follow the FCM scheme and assume that at the lower (local) and upper (global) level a single fixed value of the fuzzification coefficient "m" is considered.

Passive approach. Formation of granular prototypes at the higher level of hierarchy is reflective of the diversity of results of information granulation done at the lower level (see Figure 5.19). The information granules already formed at the lower level are not modified. In a nutshell, we are concerned with the description of the locally available information granules and the globally formed results characterize the existing variability.

Having formed information granules at the lower level, which are described by prototypes $\{v_i[ii]\}$, $i = 1, 2,..., c_{ii}$, $ii = 1, 2, ..., p$, these prototypes are treated as data to be clustered (together we have $c_1 + c_2 +...+ c_p$ elements). As a result "c" information granules are formed and characterized (described) by the prototypes $v_1, v_2, ...,v_c$. These prototypes obtained globally are evaluated vis-à-vis the local prototypes present at the lower level of the hierarchy. To do this, the links among the prototype v_i and the most consistent, highly activated prototypes formed at the lower level are established (see Figure 5.19b). Proceeding with the details, one determines a degree of membership of $v_j[ii]$ to the prototypes $v_1, v_2, ...,v_c$ using the expression

$$\xi_{i,j}[ii] = \frac{1}{\sum_{l=1}^{c} \left(\frac{\| v_i - v_j[ii] \|}{\| v_l - v_j[ii] \|} \right)^{2/(m-1)}} \tag{5.33}$$

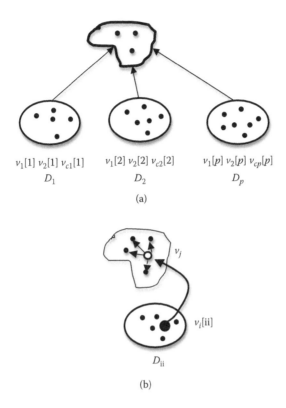

$v_1[1]\ v_2[1]\ v_{c1}[1]$ $v_1[2]\ v_2[2]\ v_{c2}[2]$ $v_1[p]\ v_2[p]\ v_{cp}[p]$

D_1 D_2 D_p

(a)

v_j

$v_i[ii]$

D_{ii}

(b)

FIGURE 5.19

Formation of higher-type information granularity—granular prototypes: a passive approach. (a) A general scheme. (b) Formation of granular prototypes V_i.

$i = 1, 2, \ldots, c; j = 1, 2, \ldots, c_{ii}$. On this basis, one determines the prototype in $D[ii]$, which is the closest to vj[ii], that is, it is characterized by the highest value of the matching level. We determine the index $j_0(i, [ii])$ for which such maximal value is attained,

$$j_0(i, [ii]) = \arg\max_{j=1,2,\ldots,cii} \xi_{i,j}[ii] \tag{5.34}$$

For v_i, we form a family of "p" prototypes at the lower level of the hierarchy, which come with the highest values of $j_0(i, [ii])$, $ii = 1, 2, \ldots, p$, that is, $v_{j0(i,1)}$, $v_{j0(i,2)}, \ldots, v_{j0(i,p)}$.

The prototype(s) identified in this way for each data are then used to form a granular representation of the prototype v_i giving rise to information granule (hyperbox or fuzzy set) V_i. A certain modification to the above process could be sought: several prototypes at D_{ii} associated with v_i to the highest values could be selected.

Active approach. The results locally available are assessed in the presence of global information prototypes. The flow of interaction is outlined in Figure 5.20.

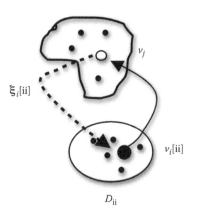

FIGURE 5.20
Interaction links formed between the levels of hierarchy.

The prototype $v_i[ii]$ coming from D_{ii} is passed on to the upper level of hierarchy and here the degrees of matching with $v_1, v_2, \dots, v_c$ are determined by using the expression whose structure is similar to the one of Equation (5.33),

$$\eta_j[ii] = \frac{1}{\sum_{l=1}^{c}\left(\frac{\|v_i[ii]-v_j\|}{\|v_i[ii]-v_l\|}\right)^{2/(m-1)}}$$

(5.35)

Computing Equation (5.35) for all locally available prototypes, we determine their compatibility with the global structure. On the basis of these compatibility (matching) degrees $\eta_{i,j}[ii]$, we form a feedback mechanism using the local structure, which could be adjusted to make it more in line with the general findings produced at the higher level. In more detail, the feedback link for the *i*-th information granule at the lower level can be determined in several ways:

(a) Optimistic. We use $\eta_j[ii]$ equal to $\max_{i=1,2,\dots,c} \eta_{i,j}[ii]$
(b) Pessimistic. $\eta_j[ii]$ is set as $\min_{i=1,2,\dots,c} \eta_{i,j}[ii]$
(c) Aggregate (average). $\eta_j[ii]$ is taken as some aggregation Φ of the matching levels $\Phi \{\eta_{1,j}[ii], \eta_{2,j}[ii],\dots, \eta_{c,j}[ii]\}$

Choosing one of the options above, we compute $w_j[ii] = 1 - \eta_j[ii]$ which quantifies a feedback level used in further adjustments of the clusters and results in a modified objective function in the following form

$$Q = \sum_{j=1}^{c_{ii}}\sum_{k=1}^{N_{ii}} w_j[ii]u_{jk}^m[ii]\, \| x_k - v_j[ii]\|^2$$

(5.36)

The weight $w_j[ii]$ impacts the calculations of the partition matrix. If $w_i[ii]$ is equal to 1 meaning that a perfect match has been reported, the corresponding row of the partition matrix is not changed in comparison with the preliminary version (as noted when discussing the iterative process of optimization). The weights do not explicitly appear in the formulas for the prototypes. Those are indirectly affected by the updated partition matrix.

The process of adjusting the local structures is realized in an iterative fashion as outlined below:

Start with the determination of the local structures without any feedback by running (locally) the FCM algorithm,
Repeat
　　Obtain the feedback links for all $ii = 1, 2, ..., p$
　　Minimize Equation (5.36) by forming updated prototypes and partition matrices
Until the process has converged (no further changes to the prototypes are reported).

5.6 Conclusions

The principle of justifiable granularity and clustering algorithms offer two different ways of forming information granules. Both of them come with a significant level of generality, however, there is a significant difference between them: clustering forms a collection of information granules while the principle of justifiable granularity leads to a formation of a single information granule. In this sense, these two techniques cannot be used interchangeably, however, they can be used in a cooperative fashion. For instance, once fuzzy clustering has been formed, one can construct granular prototypes by invoking the principle of justifiable granularity. In this construct, the data are weighted by the membership grades obtained through the clustering procedure.

References

Bezdek, J. 1981. *Pattern Recognition with Fuzzy Objective Function Algorithms.* New York: Plenum Press.

Dubois, D. and H. Prade. 1980. *Fuzzy Sets and Systems: Theory and Applications.* New York: Academic Press.

Hathaway, R.J., J. Bezdek, and W. Pedrycz. 1996. A parametric model for fusing heterogeneous fuzzy data. *IEEE Transactions on Fuzzy Systems*, 4, 270–281.

Loia, V., W. Pedrycz, and S. Senatore. 2007. Semantic web content analysis: A study in proximity-based collaborative clustering. *IEEE Transactions on Fuzzy Systems*, 15, 6, 1294–1312.

Pawlak, Z. 1982. Rough sets. *Int. J. Comput. Inform. Sci.*, 11, 341–356.

Pedrycz, W. 2005. *Knowledge-Based Clustering: From Data to Information Granules.* Hoboken, NJ: J. Wiley.

Pedrycz, W. 2009. From fuzzy sets to shadowed sets: Interpretation and computing. *Int. J. of Intelligent Systems*, 24, 1, 48–61.

Pedrycz, W., J. Bezdek, R.J. Hathaway, and W. Rogers. 1998. Two nonparametric models for fusing heterogeneous fuzzy data. *IEEE Transactions on Fuzzy Systems*, 6, 411–425.

Pedrycz, W. and F. Gomide. 1998. *An Introduction to Fuzzy Sets: Analysis and Design.* Cambridge: MIT Press.

Pedrycz, W. and P. Rai. 2009. A multifaceted perspective at data analysis: A study in collaborative intelligent agents systems. *IEEE Transactions on Systems, Man, and Cybernetics, Part B: Cybernetics*, 39, 4, 834–844.

Pedrycz, W. and J. Waletzky. 1997. Fuzzy clustering with partial supervision. *IEEE Trans. on Systems, Man, and Cybernetics*, 5, 787–795.

6

Optimal Allocation of Information Granularity: Building Granular Mappings

Information granularity helps achieve better rapport with reality by bringing into the picture the issue of nonnumeric data or results and quantifying their nature via information granules. This aspect becomes especially clear and visible in system modeling. There are no ideal models. The numeric, precise outcomes produced by models are not realistic. Information granularity has been engaged in one way or another in quantifying the lack of numeric precision. One admits a certain level of information granularity to make the model reflective of reality and quantify a limited knowledge about a phenomenon the model deals with.

Information granularity can be then regarded as an essential design asset whose prudent usage becomes crucial to make models more realistic. This position gives rise to another fundamental principle of Granular Computing—an allocation of information granularity along with an optimization of information the allocation process (Pedrycz, 2012). In system modeling, an allocation of information granularity elevates the existing models, no matter what their origin is, to a new level that could be referred to as *granular* models. In this chapter, we establish a concept of allocation of information granularity regarded as an important design asset in system modeling by giving rise to granular models. Along with the concept, discussed are protocols of allocation of information granularity throughout the system and the ensuing optimization.

6.1 From Mappings and Models to Granular Mappings and Granular Models

Let us start with an illustrative example. A function f: $\mathbf{R} \to \mathbf{R}$ is reflective of experimental data, Figure 6.1. The function could be the best approximation of the data but there always are some data that cannot ideally fit this function. As a matter of fact, only a few data could satisfy the constructed function. A lot of data points are left out. A question arises as to how the function could be made more aligned with the data. An evident possibility is to give up a numeric function and instead form a granular mapping (Bargiela and Pedrycz, 2005, 2008;

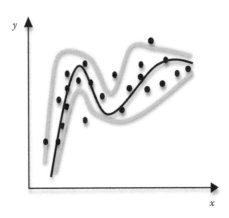

FIGURE 6.1
Function approximating experimental data and its granular generalization.

Zadeh, 1999; Pedrycz, Skowron, and Kreinovich, 2008) as shown in Figure 6.1. In this way the data become *covered* by the granular mapping (function).

Models of systems are mappings, which are constructed through system identification. Numeric models are not ideal. To make them become more aligned with the real world (experimental data), we generalize them to the form of *granular* models. There are a number of interesting and practically legitimate design and application scenarios where the inherent granularity of the models plays a visible and important role. We briefly highlight the main features of these modeling environments.

Granular characterization of models. It is needless to say that there are no ideal models, which can capture the data without any modeling error meaning that the output of the model is equal to the output data for all inputs forming the training data. To quantify this lack of accuracy, we give up on the precise numeric model (no matter what particular format it could assume) and make the model granular by admitting granular parameters and allocating a predetermined level of granularity to the respective parameters so that the granular model obtained in this way cover as many training data as possible.

Emergence of granular models as a manifestation of transfer knowledge. Let us consider that for a current problem at hand we are provided with a very limited data set—some experimental evidence (data) **D** expressed in terms of input–output pairs. Given this small data, two possible scenarios could be envisioned:

(a) We can attempt to construct a model based on the data. As the current data set is very limited, designing a new model does not look quite feasible: it is very likely that the model cannot be constructed at all, or even if formed, the resulting construct would be of low quality.

(b) We would like to rely on the existing model, which (although it deals with not the same situation) has been formed on a large and quite representative body of experimental evidence. We may take advantage of the experience accumulated so far and augment it in a certain sense so that it becomes adjusted to the current albeit quite limited data. In doing this, we fully acknowledge that the existing source of knowledge has to be taken with a big grain of salt and the outcomes of the model have to be reflective of the partial relevance of the model in the current situation. We quantify this effect by making the parameters of the model granular (i.e., more abstract and general) so that one can build the model around the conceptual skeleton provided so far. In this case, viewing the model obtained so far as a sound source of knowledge, we are concerned with the concept of an effective knowledge transfer (see Figure 6.2). The knowledge transfer (which, in essence, is represented by some model) manifests in the formation of a more abstract version of the original model.

Granular models as a result of model reduction. Models, especially those with a modular structure can give rise to their granular counterparts. This could be a result of model reduction: the reduced structure of the model is made granular, which we use to compensate for this reduction. This effect is visible quite vividly in the case of rule-based models or fuzzy rule-based models. Let us recall that those are the models composed by "P" rules of the form

$$\text{if condition is } A_i \text{ then conclusion is } B_i \tag{6.1}$$

$i = 1, 2, \ldots, P$. Now if we choose only a subset of "T" rules out of the entire collection, these rules are made granular by making the condition part A_i granular, schematically denoted as $G(A_i)$. In other words, we compensate for the reduction of the number of rules by making the remaining ones granular. These rules take on the following form

$$\text{-if } G(A_i) \text{ then } B_i \tag{6.2}$$

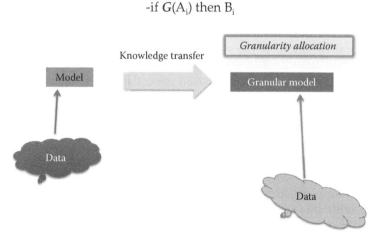

FIGURE 6.2
Emergence of a granular model as a result of knowledge transfer.

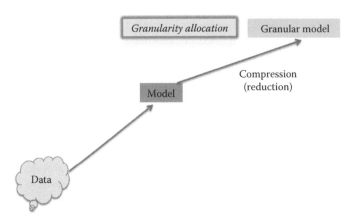

FIGURE 6.3
From a rule-based model to its reduced granular rule-based structure.

i = 1, 2, ..., T, T<<P. The essence of the reduction process is illustrated in Figure 6.3.

The choice of the subset of the rules along with an allocation of information granularity is subject to optimization. The objective of this optimization is to make the outputs of the granular model (rule-based system) as close as possible to the outputs produced by the complete model (all rules). For the predetermined subset of rules, we are concerned with a suitable distribution of information granularity among the condition parts of the already selected rules.

Granular model in modeling of nonstationary phenomena. A model of a nonstationary system is affected by the temporal changes of the system. Instead of making continuous updates to the model, which may result in a significant development overhead, one could admit a granular model with granular parameters. The granular form of the parameters is used here to account for the temporal variations of the system. In a nutshell, one constructs a model over a certain time-limited window and generalizes its numeric parameters to the granular counterparts based upon the data available outside the window (see Figure 6.4). In some sense this concept corresponds with the concept of the knowledge transfer, however the organization of the data is different than encountered there.

The mapping itself can be realized in various ways depending upon its original realization and a way in which information granules are represented; we come up with a plethora of modeling constructs with some representative examples listed in Table 6.1.

Along with this table, we can emphasize the generality and diversity of granular mappings (models) by visualizing the emerging hierarchy in Figure 6.5.

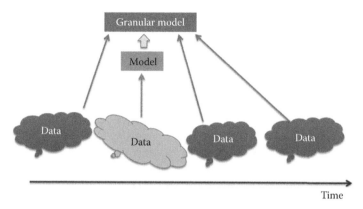

FIGURE 6.4
A granular model as a manifestation of modeling of nonstationary system.

TABLE 6.1

A Collection of Selected Examples of Granular Mappings Developed on the Basis of Well-Known Numeric Modeling Constructs

Model	Granular Model	Examples of Granular Models
Linear regression	Granular linear regression	Fuzzy linear regression Rough linear regression Interval-valued linear regression Probabilistic linear regression
Rule-based model	Granular rule-based model	Fuzzy rule-based model Rough rule-based model Interval-valued rule-based model Probabilistic rule-based model
Fuzzy model	Granular fuzzy model	Fuzzy fuzzy model = fuzzy^2 model Rough fuzzy model Interval-valued fuzzy model Probabilistic fuzzy model
Neural network	Granular neural network	Fuzzy neural network Rough neural network Interval-valued neural network Probabilistic neural network
Polynomial	Granular polynomial	Fuzzy polynomial Rough polynomial Interval-valued polynomial Probabilistic polynomial

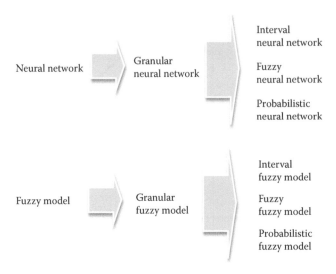

FIGURE 6.5
A plethora of granular mappings. Shown are examples of formalisms of information granularity applied to two selected categories of the mappings.

6.2 Granular Mappings

Consider a certain mapping $y = f(x, a)$ with a being a vector of parameters of the mapping. The granulation mechanism G is applied to the parameters a giving rise to its granular counterpart, $A = G(a)$ and subsequently producing a granular mapping, $Y = G(f(x,a)) = f(x, G(a)) = f(x, A)$. To form the granular mapping, we are provided with a certain level of information granularity $\varepsilon \in [0,1]$. As it helps us arrive at a more general form of the mapping and make the mapping more flexible, we can think of information granularity as an important design asset. Having this asset available, we transform the vector of numeric parameters a into a vector whose coordinates are information granules $A = [A_1 \ A_2 \ ... \ A_p]$ such that the level of admissible granularity ε is allocated to A_is in such a way that a balance of levels of information granularity with $\varepsilon_1 \ \varepsilon_2 \ ... \ \varepsilon_p$ being the levels of information granularity is satisfied, that is, $\sum_{i=1}^{p} \varepsilon_i = p\varepsilon$ i.e., $\varepsilon = \sum_{i=1}^{p}\varepsilon_i/p$. Concisely, we can capture the essence of the process of granularity allocation as follows:

$$f(x, a) \to \text{granularity allocation } (\varepsilon) \to f(x, A) = f(x, G(a))$$

numeric mapping　　　　　　　　　　　　　*granular* mapping　　(6.3)

(see also Figure 6.6). $A_i = G(a_i)$ with $G(.)$ denotes a transformation of the numeric parameter a_i to a certain granular counterpart A_i. Note that this expression is general and we are not confined to any particular formalism of information granularity used in this construct.

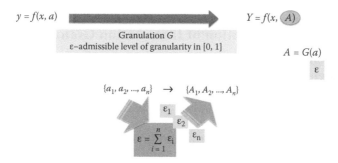

FIGURE 6.6
From numeric to granular mapping realized through the optimal allocation of information granularity.

The diversity of the granular mappings is illustrated in Figure 6.7 where we highlight the specific form of the granular output depending upon the formalism being studied (Zadeh, 1997). With this regard, it is worth noting that the granular mapping is a relation (as for any input there is a collection of outputs, namely, an information granule).

In the realization of any granular mapping, two fundamental questions arise:

- How to allocate (distribute) the given level of information granularity to the parameters of the original mapping, and
- How to optimize this process, namely, how to measure how good the constructed granular mapping is.

The two issues will be discussed in the subsequent sections. We start with outlining several ways of distributing information granularity by introducing protocols of allocation of information granularity and then look at the optimization criteria.

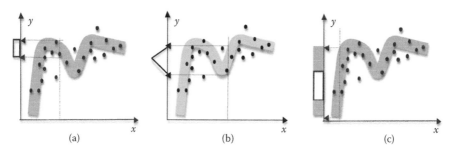

FIGURE 6.7
Examples of granular mappings highlighting the form of the information granules being generated. (a) Intervals. (b) Fuzzy sets. (c) Shadowed set.

6.3 Protocols of Allocation of Information Granularity

An allocation of the available information granularity can be realized in several different ways depending on how much diversity one would like to consider in the allocation process. In what follows, we discuss several main protocols of allocation of information granularity (refer also to Figure 6.8):

P_1: Uniform allocation of information granularity. This protocol is the simplest one. It does not call for any optimization. All numeric values of the parameters are treated in the same way and become replaced by intervals of the same length. Furthermore, the intervals are distributed symmetrically around the original values of the parameters.

P_2: Uniform allocation of information granularity with asymmetric position of intervals around the numeric parameter. Here we encounter some level of flexibility: even though the intervals are of the same length, their asymmetric localization brings a certain level of flexibility, which could be taken advantage of during the optimization process. More specifically, we allocate the intervals of lengths $\varepsilon\gamma$ and $\varepsilon(1-\gamma)$ to the left and to the right from the numeric parameter where $\gamma \in [0,1]$ controls asymmetry of localization of the interval whose overall length is ε. Another variant of the method increases an available level of flexibility by allowing for different asymmetric localizations of the intervals that can vary from one parameter to

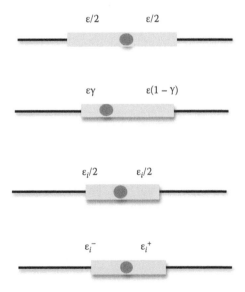

FIGURE 6.8
Protocols of allocation of information granularity P_1–P_4 and the resulting granular realization of the fuzzy sets of condition.

another. Instead of a single parameter of asymmetry (γ), we admit individual γ_i for various numeric parameters.

P_3: Nonuniform allocation of information granularity with symmetrically distributed intervals of information granules.

P_4: Nonuniform allocation of information granularity with asymmetrically distributed intervals of information granules. Among all the protocols discussed so far, this one exhibits the highest level of flexibility.

P_5: An interesting point of reference, which is helpful in assessing a relative performance of the above methods, is to consider a random allocation of granularity. By doing this, one can quantify how the optimized and carefully thought out process of granularity allocation is superior over a purely random allocation process.

In all these protocols, we ensure that the allocated information granularity meets the constraint of the total granularity available, that is, εp. (Recall that "p" denotes the number of parameters of the mapping.)

No matter whether we are considering swarm optimization (i.e., particle swarm optimization, PSO) or evolutionary techniques (including genetic algorithms, evolutionary algorithms, or the like), the respective protocols imply the certain content of a particle or a chromosome. The length of the corresponding string depends upon the protocol, which becomes longer with the increased specialization of granularity allocation.

Having considered all components that in essence constitute the environment for allocation of information granularity, we can bring them together to articulate a formal optimization process.

6.4 Design Criteria Guiding the Realization of the Protocols for Allocation of Information Granularity

Considering possible ways of allocating granularity and in order to arrive at its optimization throughout the mapping, we have to translate the allocation problem to a certain optimization task with a well-defined performance index and the ensuing optimization framework. In the evaluation, we use a collection of input–output data $\{(x_1, target_1), (x_2, target_2)\ldots (x_N, target_N)\}$. For x_k, the granular mapping return Y_k, $Y_k = f(x_k, A)$. There are two criteria of interest which are afterward used to guide the optimization of the allocation of information granularity:

(a) *Coverage criterion.* We count the number of cases when Y_k covers target$_k$. In other words, one can engage a certain inclusion measure, that is, incl (target$_k$, Y_k) quantifying an extent to which target$_k$ is included in Y_k. The computing details depend upon the nature of the information granule Y_k. If Y_k is an interval then the measure returns 1 if $\text{target}_k \subset Y_k$. In case Y_k is a fuzzy set, the inclusion measure returns $Y_k(\text{target}_k)$, which is a membership degree of target_k in Y_k. The overall coverage criterion is taken as a sum of degrees of inclusions for all data relative to all data, namely,

$$Q = \frac{1}{N} \sum_{k=1}^{N} \text{incl}(\text{target}_k, Y_k) \qquad (6.4)$$

(b) *Specificity criterion.* Here our interest is in quantifying the specificity of the information granules $Y_1, Y_2,..., Y_N$. A simple alternative using the f measure could be an average length of the intervals $V = 1/N \sum_{k=1}^{N} |y_k^+ - y_k^-|$ in the case of interval-valued formalism of information granules, $Y_k = [y_k^-, y_k^+]$ or a weighted length of fuzzy sets when this formalism is used.

Two optimization problems are formulated:

Maximization of the coverage criterion. Maximize Q realized with respect to allocation of information granularity ε, that is,

$$\text{Max}_{\varepsilon_1, \varepsilon_2 ... \varepsilon_p} Q$$

subject to constraints,

$$\varepsilon_i > 0$$

and the overall level of information granularity requirement

$$\sum_{i=1}^{p} \varepsilon_i = p\varepsilon \qquad (6.5)$$

Minimize average length of intervals. V,

$$\text{Min}_{\varepsilon_1, \varepsilon_2 ... \varepsilon_p} V$$

subject to constraints,

$$\varepsilon_i > 0$$

and the overall level of information granularity requirement

$$\sum_{i=1}^{p} \varepsilon_i = p\varepsilon \qquad (6.6)$$

This optimization is about the maximization of specificity of the granular mapping (quantified by the specificity of the output of the mapping). Note that both Q and V depend upon the predetermined value of ε. Evidently Q is a nondecreasing function of ε. If the maximization of Q is sought, the problem can be solved for each prespecified value of ε and an overall performance of the granular mapping can be quantified by aggregation over all levels of information granularity, namely,

$$AUC = \int_0^1 Q(\varepsilon) d\varepsilon \tag{6.7}$$

which is nothing but an area under curve (AUC) (see Figure 6.9). The higher the AUC value, the higher the overall performance of the granular mapping.

The criteria of coverage and the specificity of the granular outputs are in conflict. One can also consider a two-objective optimization problem and as a result develop a Pareto front of nondominated solutions.

6.5 Granular Neural Networks as Examples of Granular Nonlinear Mappings

As a representative example of the mapping, we look at fuzzy neural networks. They are constructs of computational intelligence formed at the junction of the technologies of fuzzy sets and neurocomputing (Gobi and Pedrycz, 2007; Liang and Pedrycz, 2009). The functional components of the network–logic neurons belong to two main categories of so-called AND and OR neurons (see Figure 6.10).

The OR neuron realizes an *and* logic aggregation of inputs $\mathbf{x} = [x_1 \, x_2 ... x_n]$ with the corresponding connections (weights) $\mathbf{w} = [w_1 \, w_2 ... \, w_n]$ and then

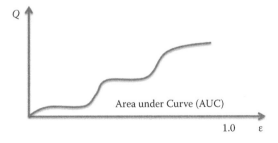

FIGURE 6.9
Performance index as a function of the level of granularity ε with an area under curve (AUC) regarded as a global descriptor of the quality of the granular model.

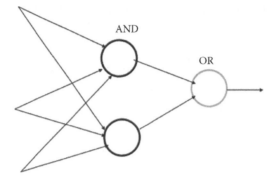

FIGURE 6.10
An example fuzzy neural network formed with the use of logic AND and OR neurons.

summarizes the partial results in an *or*-wise manner (hence the name of the neuron). The concise notation underlines this flow of computing, y = OR(x; **w**) while the realization of the logic operations gives rise to the expression (commonly referring to it as an s-t combination or more generally, an s-t aggregation of the inputs and the corresponding connections)

$$y = \overset{n}{\underset{i=1}{S}}(w_i t x_i) \tag{6.8}$$

T-norms and t-conorms (s-norms) are the generic models of logic operators used in fuzzy sets. Lower values of w_i discount the impact of the corresponding inputs; higher values of the connections (especially those positioned close to 1) do not affect the original truth values of the inputs resulting in the logic formula. In limit, if all connections w_i, $i = 1, 2,\ldots,n$ are set to 1 then the neuron produces a plain *or*-combination of the inputs, $y = x_1$ *or* x_2 *or* $\ldots$ *or* x_n. The values of the connections set to zero eliminate the corresponding inputs. Computationally, the OR neuron exhibits nonlinear characteristics (that is, it is inherently implied by the use of the t-norms and t-conorms, which are evidently nonlinear mappings). The connections of the neuron contribute to its adaptive character; the changes in their values form the crux of the parametric learning. AND neurons are described as y = AND(x; **w**) with x and **w** being defined as in case of the OR neuron, and are governed by the expression

$$y = \overset{n}{\underset{i=1}{T}}(w_i s x_i) \tag{6.9}$$

Here the *or* and *and* connectives are used in a reversed order: first the inputs are combined with the use of the t-conorm (s-norm) and the partial

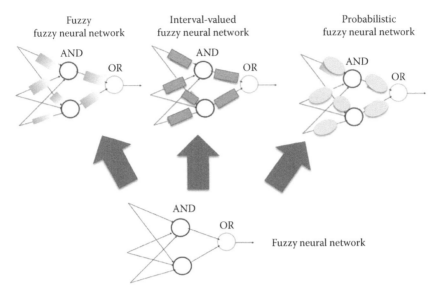

FIGURE 6.11
From fuzzy neural networks to granular fuzzy neural networks. Shown are selected realizations of information granularity.

results produced in this way are aggregated *and*-wise. Higher values of the connections reduce impact of the corresponding inputs. In limit $w_i = 1$ eliminates the relevance of x_i. With all connections w_i set to 0, the output of the AND neuron is just an *and* aggregation of the inputs $y = x_1$ *and* x_2 *and* ... *and* x_n. They also come with a significant level of plasticity whose usage becomes beneficial when learning the networks including such neurons.

By making the connections granular and admitting a certain formalism of information granularity, we end up with *granular* fuzzy neural networks such as interval-valued, fuzzy, and probabilistic fuzzy neural networks (see Figure 6.11).

In what follows, we discuss one of the granular realizations where the connections are made interval valued. We consider the detailed example shown in Figure 6.12 where a fuzzy neural network exhibits a hidden layer comprising two AND neurons. The connections of the network are also indicated in the figure. The data set **D** consists of 100 inputs x_k distributed uniformly in the $[0,1]^4$ hypercube. The corresponding outputs are affected by noise and these input–output data are used in the optimization of allocation of information granularity. The results obtained for the five protocols are visualized in Figure 6.13.

The plots of the coverage versus the levels of information granularity ε show the increased level of sophistication utilized to allocate granularity results in a visible improvement of the coverage of outputs of the data. As expected, the random allocation performs quite poorly. The uniform

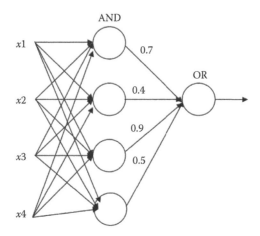

FIGURE 6.12
A fuzzy neural network.

distribution where each connection is affected to the same degree, no matter whether symmetric or asymmetric location of the interval is considered, is not very beneficial. The improvement happens when the allocation of granularity has been optimized; the advantages of the PSO are clearly evident. Likewise asymmetric position of intervals of the connections results in further improvements of the coverage. The AUC values quantify the obtained

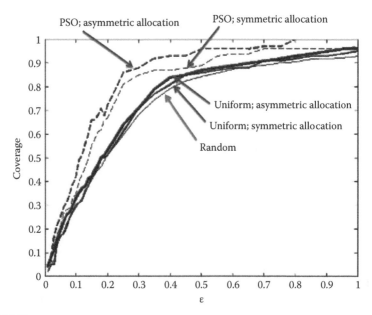

FIGURE 6.13
Coverage as a function of level of coverage ε for different protocols of allocation of information granularity.

performance. We have obtained the following results of granularity allocation: random: 0.719, uniform, symmetric allocation: 0.731, uniform, asymmetric allocation: 0.747, PSO, symmetric allocation: 0.801, PSO, asymmetric
allocation: 0.842.

It is apparent and not surprising that the two protocols in which PSO has
been used produce the best results.

With the two-criteria optimization considered (that is coverage and average length of the output information granules), we obtain a Pareto front
(Dembczynski, Greco, and Slowinski, 2009; Greco, Matarazzo, and Slowinski,
2007) as illustrated in Figure 6.14. The criteria of coverage (the plot shows
its complement, 1-coverage) and average length are in conflict and the front
is helpful in locating a sound point of compromise. The result of a single-
objective optimization (circle) is also shown.

6.6 Further Problems of Optimal Allocation
of Information Granularity

The problem of allocation of information granularity discussed so far is a
generic one, however, we can look at some other formulations, which also
come with well-justified and compelling reasons.

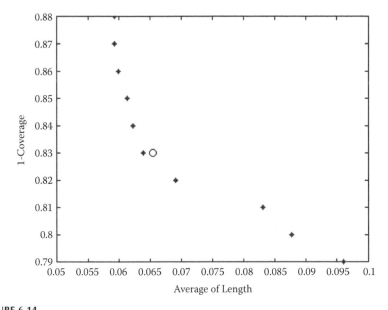

FIGURE 6.14
A Pareto front of solutions in (1-coverage, average length) coordinates. The value of ε is
equal to 0.03.

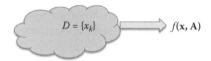

FIGURE 6.15
Allocation of granularity throughout the parameters of the mapping.

6.6.1 Specificity Maximization through Allocation of Information Granularity

An allocation of information granularity can be realized with respect to the parameters of the mapping but in a different scenario as presented so far. Here, in contrast to the previous formulation of the problem, we are provided with the inputs and make the parameters of the mapping granular, Figure 6.15, so that the allocation of granularity is realized in a way the granular output of the mapping $f(\mathbf{x}, \mathbf{A})$ is made the most specific result. Note that in comparison with the previous optimization problem where we considered a coverage criterion because of the required output (target), here there is no target value.

The problem formulated in this way can be viewed as a generalized version of the sensitivity analysis of the mapping: we are interested how different parameters of the mapping can be modified (granulated) with a minimal impact on the reduced specificity of the outputs of the mapping. The parameters whose granular counterpart comes with the lowest specificity (could be replaced by broad intervals) are those that exhibit the lowest sensitivity and the precise estimation of their values is not critical. Likewise the requirement on any possible hardware realization of the mapping could be also relaxed.

6.6.2 Optimal Allocation of Granularity in the Input Space: A Construction of the Granular Input Space

The scheme of allocation of granularity visualized in Figure 6.16 is concerned with allocation of granularity among the input variables of the mapping.

The objective is to make input variables granular instead of numeric. From a formal point of view, we arrive at the optimization problem where we maximize specificity (granularity) of the granular output of the mapping (because of the granular input variables considered here) for a predetermined

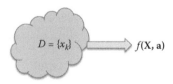

FIGURE 6.16
Allocation of granularity associated with input variables of the mapping.

level of information granularity. The ensuing optimization problem comes in the form

$$\text{Min}_{\varepsilon_1, \varepsilon_2 \ldots \varepsilon_n} \sum_{k=1}^{N} |y_k^+ - y_k^-| \tag{6.10}$$

where, as before, x_k comes from some predetermined collection of input data $\mathbf{D}$ and $Y_k = [y_k^-, y_k^+] = f(x_k, \mathbf{A})$. The allocation of granularity where each input variable becomes affected is aimed at the maximization of granularity of the output.

The problem formulated in this way helps identify variables, which are associated with the lowest level of granularity (the variables with broad intervals)—those are the variables whose values need not be specified in a precise way. In this sense, we gain a better insight as to the ranking of input variables with respect to their precision. In particular, in decision-making models when the values of inputs have to be determined and this requires some careful estimation in the presence of limited resources, the results presented here help assign effort to obtain the estimates of the inputs.

These two scenarios presented above could be arranged together by admitting information granularity to be allocated to the inputs as well as the parameters of the mapping (see Figure 6.17). This brings the joint problems of determining the sensitivity of the mapping (with respect to its parameters) and the formation of its granular inputs. While the previous way of dealing with information granularity generalizes the problem of analyzing sensitivity of "f" with regard to its parameters, in this formulation of the problem, we look at the sensitivity expressed both in terms of the parameters and the input variables.

6.7 Conclusions

At the current stage of development in Granular Computing, we urgently need a unified view and some general paradigms, independent from the diversity of formal and quite diverse settings of information granules and

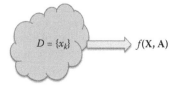

FIGURE 6.17
Allocation of granularity to parameters of the mapping and its input variables.

processing information granules. The treatment of information granularity regarded as an essential design asset of system modeling opens a new direction in system modeling by supporting a design of granular models, which are a new category of modeling constructs raising the existing (numeric) models to the next level of abstraction. The constructs introduced here were discussed in detail (along with illustrative examples) with the use of interval-based information granules. This was done on purpose as interval information granules come with transparent interpretations not cluttered with detailed computing. It has to be stressed that any other formal setting of information granulation can be considered here as well. We provided a number of examples stemming from various faculties of system modeling that highlight the generality and usefulness of the granular constructs.

References

Bargiela, A. and W. Pedrycz. 2005. Granular mappings. *IEEE Transactions on Systems, Man, and Cybernetics, Part A*, 35, 2, 292–297.

Bargiela, A. and W. Pedrycz. 2008. Toward a theory of granular computing for human-centered information processing. *IEEE Transactions on Fuzzy Systems*, 16, 2, 320–330.

Dembczynski, K., S. Greco, and R. Slowinski. 2009. Rough set approach to multiple criteria classification with imprecise evaluations and assignments. *European Journal of Operational Research*, 198, 2, 626–636.

Gobi, A.F. and W. Pedrycz. 2007. Fuzzy modeling through logic optimization. *Int. J. of Approximate Reasoning*, 45, 488–510.

Greco, S., B. Matarazzo, and R. Slowinski. 2007. Dominance-based rough set approach as a proper way of handling graduality in rough set theory. *Transactions on Rough Sets VII*, LNCS 4400. Berlin: Springer, 36–52.

Liang, X. and W. Pedrycz. 2009. Logic-based fuzzy networks: A study in system modeling with triangular norms and uninorms. *Fuzzy Sets and Systems*, 160, 24, 3475–3502.

Pedrycz, W. Forthcoming 2012. Allocation of information granularity in optimization and decision-making models: Towards building the foundations of granular computing. *European Journal of Operational Research*.

Pedrycz, W., A. Skowron, and V. Kreinovich, eds. 2008. *Handbook of Granular Computing*. Chichester: John Wiley & Sons, 347–373.

Zadeh, L.A. 1997. Towards a theory of fuzzy information granulation and its centrality in human reasoning and fuzzy logic. *Fuzzy Sets and Systems*, 90, 111–117.

Zadeh, L.A. 1999. From computing with numbers to computing with words—From manipulation of measurements to manipulation of perceptions. *IEEE Trans. on Circuits and Systems*, 45, 105–119.

7

Granular Description of Data and Pattern Classification

Data structures, especially in classification problems, pattern recognition, and pattern classifiers impact the realization of classification algorithms, and especially induce their architectures and the nature of the classification results. Here we discuss a number of approaches in which a granular description of classification data impacts a form of the classifier. We show that the concept of information granularity plays a pivotal role in the ensuing construction of the classifiers. One may refer to the pioneering study by Bellman, Kalaba, and Zadeh (1966) pointing at the role of abstraction in pattern recognition.

7.1 Granular Description of Data—A Shadowed Sets Approach

Fuzzy clustering delivers a holistic view of the experimental data in terms of information granules represented as fuzzy sets. The membership degrees associated with individual data points are useful in quantifying a numeric level of belongingness to given information granules. This numeric quantification is definitely useful, however, in some situations one would be interested in a qualitative rather than so detailed description of membership. For instance, a rough quantification of membership could be far more descriptive and helpful in the characterization of belongingness of patterns to the clusters. Here shadowed sets or rough sets could be viewed as a useful alternative. Let us recall that the construction of rough sets is implied by a characterization of fuzzy sets in which we distinguish among full membership, full exclusion, and uncertain belongingness. Figure 7.1 illustrates the underlying concept.

Likewise, constructing shadowed sets on a basis of membership functions of the fuzzy clusters gives rise to the following regions of interest in the feature space (we assume that the cores of the shadowed sets do not overlap) (Pedrycz, 1990):

Core region. It is formed as a union of the cores of the individual clusters. In other words it comprises all elements in the feature space, which belong to one of the cores.

Uncertainty region. It embraces all elements in the feature space that belong to at least one uncertain region of the cluster and are not included in the core regions.

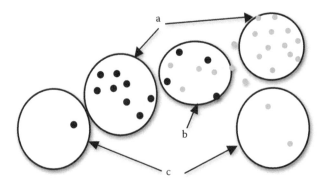

FIGURE 7.1
A three-valued view of the feature space realized by shadowed sets or rough sets for patterns belonging to four classes. The regions come with different patterns of different characteristics. (a) Regions of high homogeneity (belongingness to a single class). (b) Regions of substantial heterogeneity (patterns belonging to two or more classes). (c) Exclusion regions in which only a few patterns are located.

Exclusion region. It is composed of the elements not being positioned in the core or the uncertainty region. Typically, it is sparsely populated and comprises only a fraction of all data (patterns).

The assumption as to the nonoverlapping clusters can be easily realized by choosing a suitable number of clusters and not exceeding this number so that the non-overlap requirement can be achieved. As to the formation of the shadowed set, we follow the same principle as outlined in Section 2.5 in Chapter 2 with the only difference that now the integrals are replaced by the sums of the membership degrees.

The three-valued characterization of the feature space offers an initial qualitative view and delivers some preliminary thoughts as to the realization of classification tasks and choosing a suitable classifier depending upon the region of the feature space. For the core region, a simple classifier is sufficient as the region is highly homogeneous with patterns belonging to a single class. In the uncertainty region, a more sophisticated classifier is to be considered and designed, as it is very likely that there are patterns coming from different classes. The exclusion region might be quite homogeneous with regard to class allocation of the patterns, however, as there are few patterns/data in this region, this might lower the credibility of any classification decision issued by the classifier.

7.2 Building Granular Representatives of Data

Given is a collection of patterns, time series, multivariable data $\mathbf{X} = \{\mathbf{x}_1, \mathbf{x}_2, \ldots, \mathbf{x}_N\}$ expressed in $\mathbf{R}^n$ where $\dim(\mathbf{x}_k) = n$, thus each pattern is an n-dimensional vector of the form $\mathbf{x}_k = [x_{k1}\ x_{k2}\ \ldots x_{kn}]^T$.

To determine the *best* representative of **X**, we can consider a mean (average) of the patterns (signals) or some other statistical representative, like a median or medoid. The representatives of this nature are a result of solving an underlying optimization problem. For instance, the mean is a result of minimizing a Euclidean distance between the signals and their representative. The median results as a solution to the same problem in which the distance is specified as the Hamming one. In spite of the genuine diversity of possibilities for choosing the representative of **X**, all of these variants share a striking resemblance. The representative obtained is just an element in the same space (feature space) in which the original signals were expressed. Thus, if **X** is expressed in $\mathbf{R}^n$, so is the space in which the representative of **X** becomes formed. Formally speaking, the representation problem gives rise to the formation of the representative of **X**, call it v, such that it represents (or approximates) all elements in this data set. We can capture the essence of this category of the signal representation problems in the following manner:

$$X = \{x_1, x_2, \ldots, x_N\}, x_k \in \mathbf{R}^n \rightarrow \mathbf{v} \in \mathbf{R}^n \tag{7.1}$$

As intuition may suggest, by noting an inherent many-to-one nature of the mapping (many elements in **X** and a single representative) and in order to accommodate the diversity of the signals to be represented, one could envision that the structural complexity (a level of abstraction) of **v** is supposed to be higher than the original signals it has to represent. This entails that rather than being a vector of numeric entities, one may anticipate that the representative can be sought as a certain information granule being of nonnumeric character. For instance, we may envision that such representative could be a collection of intervals or a family of fuzzy sets formed over **R**.

The granularity of information, which is inherently associated with the representative of the signals is fully reflective of the many-to-one nature of the mapping of the elements of **X** to a single representative. More generally, we can envision the representative to be realized as any granular construct, that is, $G(\mathbf{R}^n)$. Proceeding with the concise notation conveyed by Equation (7.1), we articulate the essence of the realization of granular prototypes in the following way:

$$X = \{x_1, x_2, \ldots, x_N\}, x_k \in \mathbf{R}^n \rightarrow V \in G(\mathbf{R}^n) \tag{7.2}$$

To shed more light on the granular character of the representative, as an illustration, let us consider a simple example. In Figure 7.2a, we show four temporal signals $x_1, x_2, x_3,$ and x_4 expressed in a finite-dimensional space (so each x_k can be treated as a vector of amplitudes present in consecutive time instances). The initial representative of this group is one of the signals, around which a certain information granule is being formed. In Figure 7.2b, we show the granular realization completed in terms of intervals (sets). The effect of *covering* (namely, representing) x_k by the granular representative is apparent.

The higher diversity of the signals present at some time moments becomes reflected by the broader (less specific) information granule allocated there.

As yet another example, let us consider a family of continuous temporal signals (patterns) governed by the expression $x(t) = b*\exp(-at)$ where a and b are the two parameters of the signal. Then the signals are fully described in the two-dimensional feature space (a, b) (see Figure 7.2c). The granular representative of the signals in this two-dimensional parameter space is visualized as a box-like structure, which is nothing but a Cartesian product of the intervals formed in the individual features. More formally, we have $A \times B$ where $A = [a_-, a_+]$ and $B = [b_-, b_+]$. Note that the lengths of the sides of these boxes may not be the same. Again as above, the granular realization of the representative can be completed in the formal setting of sets (intervals) or fuzzy sets.

In both examples presented above, there is compelling evidence that the granular, rather than the numeric prototype (representative), comes with several tangible advantages. In particular, a granular representative quantifies and visualizes a diversity of signals under consideration.

The essence of the development of granular representatives can be viewed as an optimization problem of distribution of the available level of information granularity where as in many situations before, the granularity itself is treated as an important knowledge-based modeling resource. In a nutshell, given a predefined level of information granularity, we allocate it to the elements of the universe of discourse **R** in such a way that the resulting granular representative captures most of the signals (i.e., the signals are *contained* within the bounds of the information granules of the representative). The higher the admitted level of granularity ε^* is, the less specific (detailed) the granular representative becomes. This tendency is not encouraging. At the same time, with the lowered values of granularity, more data are being covered (which is evidently advantageous). These two conflicting requirements need to be carefully reconciled during the design of the granular representatives.

It is worth noting that granular representatives are of particular interest in interpreting, analyzing, and comparing diagnostic signals (e.g., electrocardiography [ECG] complexes) where the concept of a representative (norm) and its variation explicitly associated with the norm becomes of relevance. For instance, one could envision a template (representative) of normal ECG, where its granular nature augments the representation abilities of the template and reflects the variability of the existing signals and a way in which it is distributed over time. Any comparison with a new ECG complex is done with the granular template and on this basis a certain classification could be carried out.

7.2.1 A Two-Phase Formation of Granular Representatives

Let us start with formulation of the optimization problem of granularity allocation where an available level of granularity is regarded as an

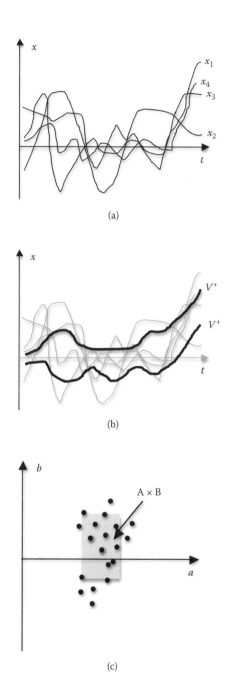

FIGURE 7.2
Examples of granular representatives. (a) Temporal signals shown in time domain. (b) The realization of the granular prototype in the form of sets and fuzzy sets. (c) Patterns in the two-dimensional feature space and their granular representation.

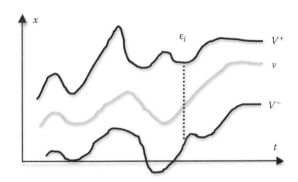

FIGURE 7.3
Formation of granular prototype around the numeric representative.

important knowledge representation resource used to help represent the entire collection of data. As before, let us consider some numeric representative $\mathbf{v}$ of the data set $\mathbf{X}$ as shown in Figure 7.3. We develop a granular extension of the prototype $\mathbf{v}$ (call it V), which is defined in $P(\mathbf{X})$. This is completed by admitting some interval-valued entries being built around the successive coordinates of $\mathbf{v}$ (see Figure 7.3). The expressive power of the information granule $\mathbf{V}$ articulated with respect to $\mathbf{X}$ is higher than the original $\mathbf{v}$ in the sense that $\mathbf{V}$ covers (includes) some entries of $\mathbf{x}_k$. The broader the interval built around the numeric representative $\mathbf{v}$, the more data points fall within the bounds of the information granule $\mathbf{V}$. Note that the length of the interval of $\mathbf{V}$ may vary over the entire space when moving from one coordinate of $\mathbf{x}_k$ to another. There are no particular restrictions on the distribution of granularity.

We translate the problem into the corresponding optimization task with the objective to allocate granularity along the universe of discourse $\mathbf{R}$ in such a way so that as many coordinates of $\mathbf{x}_k$ are included within the bounds of $\mathbf{V}$. The bounds of $\mathbf{V}$ are described in the interval-like form, that is, $\mathbf{V} = [[V_1^-, V_1^+]\,[V_2^-, V_2^+]...[V_n^-, V_n^+]]$; we are concerned with the interval type of granularity of the representative.

The optimized performance index reads as follows:

$$\text{Maximize card } \{(i, k) \,|\, x_{ki} \in [V_i^-, V_i^+]\} \tag{7.3}$$

with the maximization realized with respect to the vector of information granularity $\varepsilon = [\varepsilon_1 \; \varepsilon_2 ... \; \varepsilon_n]^T$ with the constraint imposed by the assumed cumulative level of overall granularity ε^*, that is,

$$n\varepsilon^* = \sum_{i=1}^{n} \varepsilon_i \tag{7.4}$$

The minimization is guided by the following performance index:

$$Q = N*n - \text{card}\{(i,k) \mid x_{ki} \in [V_i^-, V_i^+]\} \tag{7.5}$$

which concentrates on the maximization of the number of points located inside **V**. Obviously, the higher the value of ε^*, the more data points there are that are potentially covered by the granular representative of X.

The issue of an optimal distribution of granularity brings an important aspect of robustness of the approach. It is worth stressing that the method exhibits two interesting features making the overall method robust. First, a prototype is selected as being *one* of the elements of the collection of the signals. Second, the allocation of granularity is realized by counting the number of elements enclosed by the granular *envelope*. In this way, if we encounter an outlier, it is eliminated from the granular representation of the prototype as its possible inclusion might have required a substantial usage of the overall resource of information granularity available.

7.2.2 Optimization of Information Granularity with the Use of the Particle Swarm Optimization (PSO) Algorithm

As the PSO method is a generic optimization vehicle, to use it in the problem at hand, it becomes necessary to specify the content of the particles used in the search space. The entries of the particle **z** (located in the unit hypercube) are translated (decoded) into the granular representation of **v** by using two transformations:

- First, the entries of the particle are converted into the corresponding levels of granularity for the individual elements of the universe of discourse **X**. Recall that the level of granularity ε^* has been provided in advance. The calculations of **z** are expressed in the following way:

$$z_i^* = \frac{z_i n}{\sum\limits_{j=1}^{n} z_j} \varepsilon^* \tag{7.6}$$

- Second, the levels of granularity translate into the information granules (intervals); note that in these calculations we also involve the ranges, range_i, of the values of X assumed over the individual element of **X** for the i-th coordinate (element) of **X**. We get

$$[V_i^-, V_i^+] = [\max(x_{low}(i), v_i - z_i^*/2*\text{range}_i), \min(x_{high}(i), v_i + z_i^*/2*\text{range}_i)] \tag{7.7}$$

where

$$x_{low}(i) = \min_{k=1, 2, \ldots, N} x_{ki} \qquad x_{high}(i) = \max_{k=1, 2, \ldots, N} x_{ki} \qquad (7.8)$$

An overall flow of computing comprises two steps: (a) we pick up a certain element of X, and (b) for some given value of ε^*, we optimize a distribution of granularity so that Equation (7.5) becomes maximized. The visualization of this two-phase process is shown in Figure 7.4. Depending upon the element of X around which the granular representation is being constructed, we arrive at different values of the performance index. Likewise these values depend on some predetermined level of the granularity ε^*.

The quality of the granular representatives depends upon a choice of a certain value of the overall granularity. To avoid being potentially affected by any particular selection of this level, we introduce a global way of expressing the quality of a certain granular representative V. It is intuitively straightforward that the relationship $Q = Q(\varepsilon^*)$ is a nondecreasing function of ε^* meaning that a higher level of granularity available for distribution can result in covering more data points in X. Instead of admitting a particular value of ε^* (whose choice is usually biased to some extent and implied by some design performance), we sweep through a range of values of ε^* starting from zero (in which case Q is typically close to zero) and move to some upper bound, that is, ε_{max}. At the same time, we record the corresponding values of Q (those are optimized values of Q for the specific value of ε^*). The resulting area under curve (AUC) serves as a viable global indicator of the suitability of the granular representative V (formed via the formation of the granular representation of v),

$$AUC = \int_{0}^{\varepsilon_{max}} Q(\varepsilon)d\varepsilon \qquad (7.9)$$

The higher the value of AUC, the better the granular representative V is and this quantification is of general character independent from the required level of granularity. In this manner, any choice of v we have

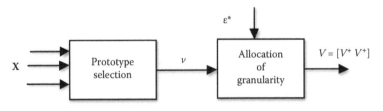

FIGURE 7.4
Optimization (distribution) of granularity in the development of granular representatives of the signals: a two-phase development process.

started with can be quantified in terms of the AUC. Alluding to the scheme illustrated in Figure 7.4, the selection of the best representative of **X** is realized by picking up its successive elements of this data set, optimizing distribution of the level of granularity, and selecting the one with the highest value of the AUC.

7.2.3 Some Illustrative Examples

In what follows, to illustrate an overall development process, we present results for some synthetic data as well as example ECG data.

Synthetic data. These data, shown in Figure 7.5, comprise four time series. The three of them (indexed by *a-b-c*) are close to each other. The fourth one (denoted by *d*) is quite distinct as its trend is opposite to the one characterizing the remaining time series.

The PSO (Kennedy and Eberhart, 1995; Van den Bergh and Engelbrecht, 2006) is realized with a population of 100 individuals and the method was run for 60 generations. The inertia weight (ξ) was set to 0.6. The PSO is realized with a population of 100 individuals and the method was run for 60 generations. The values of $Q(\varepsilon^*)$ are reported in Figure 7.6. As anticipated, the choice of the first three time series as the prototypes and their granular extension gives rise to very similar plots of $Q(\varepsilon^*)$ whereas the significant difference is observed for the fourth time series being treated as the prototype. The corresponding values of $Q(\varepsilon^*)$ are far lower than the ones observed so far. The differences are also significantly quantified by looking at the areas under curve where we obtain 40.80, 41.35, 41.15, and 26.80, respectively. Thus for the last time series, the value of the AUC amounts to about 65% of the value reported for the first three time series.

Taking as an example a certain value of $\varepsilon^* = 0.9$, we look at the plots of the granular realization of the prototypes (the second and the fourth time series), Figure 7.7. The optimal allocation of granularity is quite nonuniform.

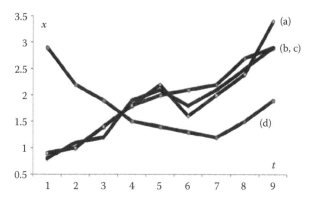

FIGURE 7.5
Synthetic data consisting of four time series.

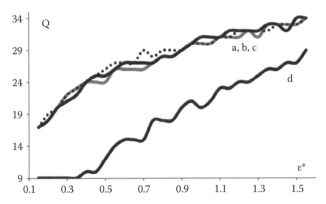

FIGURE 7.6
Performance index Q regarded as a function of ε^*.

In the case of Figure 7.7a, there are some coordinates of the prototype, which come with broad intervals (low granularity). One can relate these results with the distribution of time series. The spreads of the intervals are reflective of the numeric values of the time series. For instance, for the first coordinates the three time series coincide and this results in narrow intervals while some rippling effect is visible for the higher coordinates as the three time series start to exhibit more diversity. Furthermore, the fourth time series gets closer to the three time series and that is why it is taken into consideration when distributing granularity. Note that it has been left out for the formation of the granular prototype at the lower coordinates (simply it was too far to be included in the granular construct). The granular prototype formed on a basis of the fourth time series, Figure 7.7b exhibits a far larger variability in the lengths of the intervals because of the location of this prototype vis-à-vis the remaining time series.

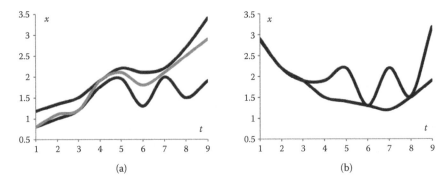

FIGURE 7.7
Granular prototypes (lower and upper bounds) for the prototypes constructed on the basis of (a) the second and (b) the fourth time series.

ECG signals. In this series of experiments, several ECG signals are used coming from the MIT-BIH arrhythmia database (MIT BIH ECG Database). All of them are represented as discrete time series centered around the R peaks of the QRS complexes. Each of the signals is of the length of 73 samples. It is assumed that the signals were preprocessed in terms of proper amplitude alignment (eliminating potential offset) and realizing eventual time warping if required.

A collection of six normal ECG complexes is shown in Figure 7.8. While these signals exhibit some similarities, there is a certain level of variability present among them. The evaluation of each of the signals in terms of the AUC measure (see Figure 7.9) indicates that the fifth one is the most suitable as a granular representative and returns the highest value of this measure with the AUC value being 18% higher than the weakest granular representative.

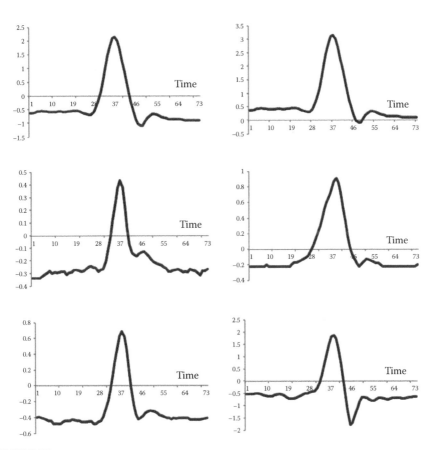

FIGURE 7.8
A collection of six normal ECG signals.

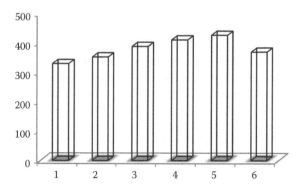

FIGURE 7.9
AUC values computed for the ECG signals.

Considering this numeric representative of the collection of the signals, we visualize its granular profile by looking at the levels of coverage versus increasing values of ε^* (see Figure 7.10). This relationship indicates a fairly steady increase of the coverage with a slight saturation effect noted for higher values of ε^*.

Considering some selected values of ε^*, that is, 0.05, 0.10, and 0.20 (those values are picked up here for illustrative purposes), the lower and upper bounds of the granular representations of the signal are illustrated in Figure 7.11.

We note that the granularity of the representative becomes more apparent with the increase of the allowed granularity level. Furthermore, the distribution of granularity (the lengths of the intervals) differs quite substantially: it is nonexistent in the neighborood of the R peak while it shows up in other regions of the QRS complex.

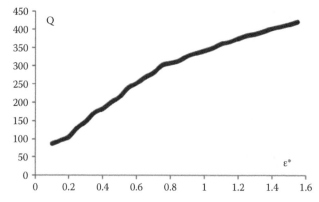

FIGURE 7.10
Number of data samples being covered versus levels of available information granularity ε^*.

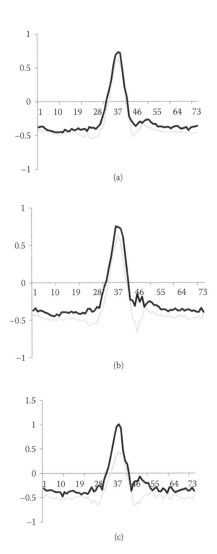

FIGURE 7.11
Granular realization of the representative for selected values of ε^*: (a) 0.05, (b) 0.10, and (c) 0.20.

7.3 A Construction of Granular Prototypes with the Use of the Granulation–Degranulation Mechanism

Adhering to the same general line of thought as discussed in Section 4.2 in Chapter 4, we develop granular prototypes on the basis of the original numeric prototypes constructed with the aid of the fuzzy c-means (FCM) method. Then they are made granular by optimizing a certain criterion based on the quality of the reconstruction obtained for such granular prototypes.

As before, we consider a set of data $X = \{x_1, x_2, ..., x_N\}$ for which is constructed a collection of "c" information granules $G_1, G_2, ..., G_c$ in the form of fuzzy sets. As usual, the prototypes $v_1, v_2, ..., v_c$ as well as partition matrices being either elements in $\{0,1\}^{c \times N}$ or $[0,1]^{c \times N}$ are plain *numeric* constructs.

Generalizing the idea of granulation–degranulation (see Section 4.2 in Chapter 4), one envisions that the granulation–degranulation may produce results that are information *granules* rather than single numeric entities. The expectations behind this construct could be that such information granules X_k produced as a result of this transformation cover (include) the original numeric input. In other words, instead of the satisfaction of the condition $\hat{x}_k = G^{-1}(G(x_k))$, we could expect a satisfaction of the following inclusion:

$$x_k \subset X_k \tag{7.10}$$

anticipating that this relationship be (hopefully) satisfied for all (or at least *most*) data x_k. The conceptual and ensuing technical question is about the origin of granularity to be used in the formation of the clusters and a way of building it in the representation of the results of clustering.

We may envisage, which is intuitively appealing, that as being more abstract than data themselves, the representatives of the granulation process can be regarded as information granules. For instance, prototypes themselves could be regarded as information granules. More formally, we consider that $V_1, V_2, ..., V_c$ are elements in the space of information granules spanned over $\mathbf{R}^n$, that is, $V_1, V_2, ..., V_c \in G(\mathbf{R}^n)$ where $G(.)$ denotes a family of information granules formed in $\mathbf{R}^n$. In contrast, let us stress that so far we assumed that the prototypes are numeric, $v_1, v_2, ..., v_c \in \mathbf{R}^n$. Following the same notation as before, the result of expressing x via information granules can be formally written as $G(x, V_1, V_2, ..., V_c, U)$. Subsequently, $D(G(x, V_1, V_2, ..., V_c, U))$ is an information granule and therefore the relationship $x_k \in D(G(x_k, V_1, V_2, ..., V_c, U))$ makes sense.

Let us highlight that in contrast to numeric prototypes, granular prototypes enhance the results of information granulation both from the point of view of the concepts as well as the perspective of applications:

- As opposed to their numeric counterparts, granular prototypes help fully describe the structure of data to be clustered. For numeric data exhibiting different levels of dispersion, the numeric prototypes could be almost the same, however, their granular augmentation is capable of telling them apart.

- The granularity of the prototypes helps distinguish between the data that are compatible with the essence of the structure of data and the data points that exhibit a certain incompatibility and might be flagged as potential outliers.

- The granularity of the prototypes can be assigned through a well-formulated optimization process. The proposed algorithm of optimal information granularity allocation stresses the role of information granularity as an essential design asset.
- The comprehensive description of data through the granular description of the prototypes augments the quality and completeness of the characterization of the structure of the data. The prototypes sought as the representatives of the data capture their dispersion in an explicit way.

7.3.1 Granulation and Degranulation Mechanisms in $G(x, V_1, V_2, ..., V_c, U)$

In the case where the descriptors of the granulation process themselves are information granules (and here V_i are elements in $G(\mathbf{R}^n)$), the mechanisms of granulation are retained the same as before (that is, we involve the numeric prototypes). In the degranulation, we take into consideration the granular nature of the prototypes. In particular, let us consider $V_1, V_2, ..., V_c$ to elements in a family of hypercubes (information granules) defined over $\mathbf{R}^n$, that is, $P(\mathbf{R}^n)$ (see Figure 7.12).

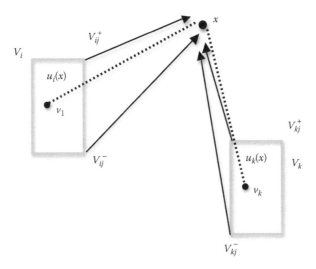

FIGURE 7.12
Granular prototypes formed around numeric representatives v_1, v_2, v_c, and their use in the realization of the granular reconstruction. Shown are prototypes V_i and V_k. Dotted lines point at the bounds of the information granules used in the calculation completed with the use of Equation (7.11).

In the degranulation, the bounds of the granular prototypes are used and they imply the granular nature of the reconstruction. Now, the degranulation returns an information granule $\hat{X}$ which is again defined in $P(R^n)$ and whose bounds are expressed as follows:

$$\hat{X} = \left[\frac{\sum_{i=1}^{c} u_i^m(x)V_i^-}{\sum_{i=1}^{c} u_i^m(x)}, \quad \frac{\sum_{i=1}^{c} u_i^m(x)V_i^+}{\sum_{i=1}^{c} u_i^m(x)} \right] \tag{7.11}$$

where $V_i = [v_i^-, v_i^+]$. When $\hat{X}$ is expressed in terms of the individual coordinates, one obtains

$$\hat{X}_j = \left[\frac{\sum_{i=1}^{c} u_i^m(x)v_{ij}^-}{\sum_{i=1}^{c} u_i^m(x)}; \quad \frac{\sum_{i=1}^{c} u_i^m(x)v_{ij}^+}{\sum_{i=1}^{c} u_i^m(x)} \right] \tag{7.12}$$

The granular prototypes (and ensuing granular fuzzy clustering) build upon fuzzy clustering (no matter what specific clustering technique has been used) and form another conceptual and algorithmic layer that augments interpretation capabilities already delivered by fuzzy clustering.

7.4 Information Granularity as a Design Asset and Its Optimal Allocation

The granularity of the result coming from the degranulation process has to have a certain origin. The granular prototypes come as a sound and justifiable alternative, however, they need to be constructed. The easiest way to form granular prototypes is to form information granules (i.e., intervals or fuzzy sets) around the prototypes obtained so far. Put it in a more descriptive way; information granularity (giving rise to granular prototypes) is an essential design asset that requires the best allocation so that the inclusion $x_k \in D(G(x_k, V_1, V_2, ..., V_c, U))$ is satisfied for as many data in X as possible. All the points x for which this relationship holds will be called *compatible* (or *granularly compatible*) with the structure formed in the clustering process.

Proceeding with the structure of the granular prototypes, V_i is built around the original prototype v_i by admitting some level of granularity ε assuming values in [0,1]. In the simplest possible scenario, we can envision the following transformation in which an information granule of the prototype is being formed

$$V_{ij} = [v_{ij} - \varepsilon * range_j, v_{ij} + \varepsilon * range_j] \tag{7.13}$$

$i = 1, 2, ..., c, j = 1, 2, ..., n$. Note that the prototype is made *granular* to the same extent with regard to all variables. All coordinates of the prototype are transformed to the intervals that are symmetrically distributed around v_{ij} and equally affected by the imposed level of granularity.

The degranulation–granulation criterion can be used as a vehicle to associate ε with the quality of the overall process. A convincing performance index Q can be introduced as the following ratio of the count of the data included in the granular realization versus all data (N):

$$Q = \frac{card\{x_k \in G^{-1}(G(x_k, V_1, V_2, ..., V_c, U))\}}{N} \tag{7.14}$$

Evidently, the higher the value of ε, the higher the value of the ratio (and the quality of the information granulation) becomes. Q is a nondecreasing function of ε. Too high values of ε might not be acceptable—so one could admit some upper limit and regard ε itself as a certain design resource used to facilitate an effective realization of the granular structure. This performance index captures the nature of consistent granular representation of data through clustering: we strive to achieve a situation where most of the patterns when being represented by the granular prototypes are positioned within the bounds resulting through the reconstruction process Equations (7.11) and (7.12).

The main limitation of the construction of the granular prototypes is that we have treated all variables in the same way by assigning to all of them the same value of ε. As granularity is sought as an important modeling asset to be prudently allocated, its distribution needs more attention. The idea of allocation of information granularity is immediately applicable here. We require that the sum of levels of granularity of V_i is equal to $n\varepsilon$. For each variable we assign some specific level of granularity ε_i such that the following balance of overall granularity resource expressed as

$$\sum_{i=1}^{n} \varepsilon_i = n\varepsilon \tag{7.15}$$

becomes satisfied. The optimization criterion to be maximized that governs this allocation is given by Equation (7.15). By noting an indirect way in which the optimization criterion depends upon the vector of nonnegative variables,

$\varepsilon = [\varepsilon_1\ \varepsilon_2\ ...\ \varepsilon_n]$ with the constraint Equation (7.13). As discussed before, PSO or genetic algorithm (GA) become a viable optimization alternative.

7.5 Design Considerations

We elaborate on the design facet of the problem, which concerns a selection of the level of allowable granularity. This directly relates to the structure description (expressed in terms of the underlying granular core) and a way of selecting a suitable level of granularity. Here we link this selection to the problem of satisfaction of a constraint of a suitable linguistic quantifier.

7.5.1 Granular Core and Granular Data Description

The union of granular prototypes V_i can be treated as a granular *core* of the data space in which the data could be reconstructed. Note that we do not require that the granular prototypes be pairwise disjoint. Considering the granular core and the implied reconstruction, we identify two categories of data:

(a) Data that are correctly degranulated (according to the condition of granular consistency, namely, a satisfaction of the inclusion $x_k \in G_{-1}(G(x_k, V_1, V_2, ..., V_c, U))$).

(b) Data that are not correctly reconstructed (for which the inclusion condition does not hold). These data can be regarded as not being *compatible* with the identified granular core.

Note that these two categories of data are indexed by the values of ε with the increase in the cardinality of the first category associated with higher values of ε.

7.5.2 Selection of a Suitable Value of Information Granularity

The relationship $I = I(\varepsilon)$ is a nondecreasing function of ε and the increasing values of ε produce higher levels of coverage of the data. While this observation is intuitively appealing, the choice of a *suitable* value of ε requires some clarification. Two ways of determining such value could be considered here:

(a) Based on the plot of I, one determines its knee-like point, call it ε_{opt}, beyond which the changes of I start to become less visible. This way of determining ε_{opt} is suitable in cases in which such point is clearly identified. Obviously, this critical point supporting the choice of ε_{opt} may not always be clearly visible and this hampers the usefulness of this criterion.

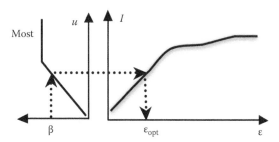

FIGURE 7.13
Determination of optimal level ε_{opt} for some predetermined satisfaction level β of the linguistic quantifier *most*.

(b) Based on the use of the linguistic quantifier $\tau = most$ (Kacprzyk and Yager, 2001; Kacprzyk and Zadrozny, 2005) where ε_{opt} is determined by seeking a level of satisfaction of the linguistic quantifier

$$most \text{ data have a consistent granular representation} > \beta$$

where β taking on values in the [0,1] interval is the predetermined level of satisfaction of the linguistic quantifier *most*. The idea of this construct is illustrated in Figure 7.13.

A way of arriving at the values of ε_{opt} that corresponds to β is through the calculations of the value of the argument of the quantifier where we have $most^{-1}(\beta) = u$ (note that the quantifier is defined over the ratio of data) and then from the relationship $I(\varepsilon) = u$ we determine ε_{opt}, that is, $I(\varepsilon_{opt}) = u$. This method is attractive as it links the less tangible values of ε with the explicit and intuitively convincing degree of satisfaction of the linguistic quantifier τ. For instance, we can resort to the linguistic expression of the form

$$most \text{ data are granularly compatible with the structure to a degree } \beta$$

where β is a predetermined level of satisfaction.

Example

As an illustrative example, we consider a two-dimensional data shown in Figure 7.14. Through the visual inspection, we distinguish three clusters. The FCM returns the following prototypes: $\mathbf{v}_1 = [1.11 \ 2.10]^T$, $\mathbf{v}_2 = [3.43 \ 2.07]^T$, and $\mathbf{v}_3 = [0.32 \ 0.25]^T$.

The *most* quantifier is modeled as (Kacprzyk and Yager, 2001)

$$\tau(u) = \begin{cases} 0, & \text{if } x \leq 0.3 \\ 2x - 0.6, & \text{if } x \in [0.3 \ 0.8] \\ 1, & \text{if } x \geq 0.8 \end{cases}$$

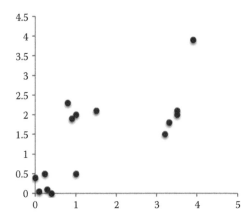

FIGURE 7.14
Two-dimensional synthetic data with three clusters.

while the threshold (satisfaction level) is set to the value $\beta = 0.8$. This implies that $u_0 = most^{-1}(0.8)$ is equal to 0.7.

The prototypes are expanded to their granular counterparts by optimizing the allocation of the available level of granularity ε. The results reported in terms of the maximized performance index I are shown in Figure 7.15. It is apparent that the PSO led to the higher values of this index in comparison with the values of I reported for the uniform allocation of granularity across all the variables. The overall improvements are quantified by the obtained values of the AUC, which are equal to 0.099 for the uniform granularity allocation and 0.108 for the PSO optimized granularity allocation.

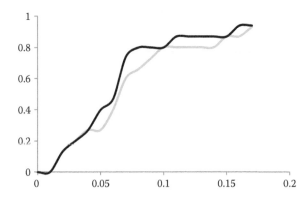

FIGURE 7.15
Plot of I versus granularity level ε. Black line: PSO-guided information granularity allocation. Gray line: uniform allocation of information granularity.

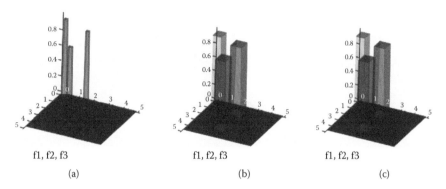

f1, f2, f3　　　　　　　f1, f2, f3　　　　　　　f1, f2, f3

(a)　　　　　　　　　(b)　　　　　　　　　(c)

FIGURE 7.16
Plots of the union of granular prototypes for selected values of ε. (a) e = 0.03 with ε = [0.028 0.031]. (b) ε = 0.1, ε = [0.108 0.09]. (c) ε = 0.16, ε = [0.173 0.146].

The plots of the core region composed of the union of X_is (as formed by (7.2)) for several selected values of ε are shown in Figure 7.16.

The data set in Figure 7.17 exhibits more scattering and less visible structure than the one in the previous example. The clustering is run for c = 3 and the obtained prototypes are given as $\mathbf{v}_1$ = [3.35 1.15], $\mathbf{v}_2$ = [3.14 3.45], and $\mathbf{v}_3$ = [0.55 1.01].

The plot of the coverage index I versus ε is displayed in Figure 7.18.

Again, as visualized there, the advantages of optimization provided by the PSO are evident. Considering the quantitative characterization provided by the AUC, we obtain AUC = 0.113 for the uniform allocation of granularity and AUC = 0.136 for the PSO-optimized allocation of granularity.

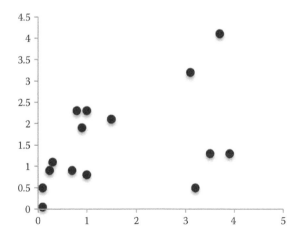

FIGURE 7.17
Two-dimensional synthetic data with more scattered patterns.

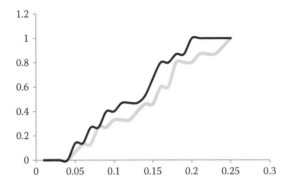

FIGURE 7.18
The values of I versus ε. Black line: PSO-guided information granularity allocation. Gray line: uniform allocation of information granularity.

7.6 Pattern Classification with Information Granules

In the plethora of pattern classifiers visible in pattern recognition (Duda et al., 2001; Kandel, 1982; Roubens, 1978; Fukunaga, 1990), there is a highly visible trend of constructing more advanced classifiers with highly non-linear architectures (being more in rapport with nonlinearity of data) and transformations of feature spaces (with support vector machines forming one of the convincing examples in this area). There is also an interesting trend to build geometric classifiers where some hyperbox regions are delineated (cf. Bargiela and Pedrycz, 2003; Pedrycz and Succi, 2005; Rizzi et al., 2000; Simpson, 1992, 1993; Bargiela and Pedrycz, 2005; Gabrys and Bargiela, 2000) or a focus is on other geometric constructs (Abe et al., 1998).

In a nutshell, however, it could be beneficial to distinguish between some regions of the feature space which are highly homogeneous and occupied by patterns coming from a single class and the other regions in this space where a far lower regularity is observed and mixtures of nonseparable patterns coming from several classes become present. Intuitively, the class homogeneous regions of the feature space occupy a significant portion of the feature space and do not call for any sophisticated architecture of the classifier as with a high level of confidence one can complete classification of patterns falling within the bounds of these regions. In contrast, in the area where we encounter a significant level of class overlap (mixture of classes), more advanced, nonlinear classifiers are required. At the end, still a significant classification error could be present there and any classification done in such regions of the feature space has to be carefully assessed and eventually treated with some level of criticism.

These observations suggest a general strategy for dealing with pattern classification: first, we start with identification of the homogeneous regions in the feature space (which could be quite easy to characterize), and then we characterize the classification process and results that concern patterns located outside the regions of high homogeneity.

Having this in mind, we come up with two well-delineated architectural components of the classification scheme, which directly reflects the functionality outlined above. The core (primary) part of the classifier, which captures the essence of the structure, is realized in terms of hyperboxes (Cartesian products of intervals in the corresponding features). We refer to it as a *core* structure. For this core structure, the use of sets as generic information granules is highly legitimate: there is no need to distinguish between these elements of the feature space as no discriminatory capabilities are required as the classification error is practically absent.

More advanced classification structures are sought outside the core structure to reflect upon the overlap among classes, the use of membership grades (invoking fuzzy sets), as well as granular membership grades. The areas of high-class overlap require more detailed treatment hence here arises a genuine need to consider fuzzy sets as the suitable granular constructs. The membership grades play an essential role in expressing levels of confidence associated with the classification result. In this way, we bring a detailed insight into the geometry of the classification problem and identify regions of very poor classification.

This leads to a two-level granular architecture of the classifier, which convincingly reflects a way in which classification processes are usually carried out: we start with a core structure and then consider the regions of high overlap between the classes where there is a high likelihood of the classification error. One can view the granular classifier as a two-level hierarchical classification scheme (Pedrycz et al., 2008) whose development adheres to the principle of a stepwise refinement of the construct with sets forming the core of the architecture and fuzzy sets forming its specialized enhancement. Given this, a schematic view of the two-level construct of the granular classifier is included in Figure 7.19.

The design of the two-level granular classifiers offers several advantages over some *standard* realizations of pattern classifiers. First, the interpretability is highly enhanced: both the structure and the conceptual organization appeals in a way in which an interpretation of the topology of patterns is carried out. Second, one can resort to the existing learning schemes developed both for set-theoretic classifiers and fuzzy classifiers.

7.7 Granular Classification Schemes

The granular nature of prototypes gives rise to a clear distinction between the regions belonging to the union of these prototypes (core region) and the remaining domain of the feature space. As noted in the previous section, the

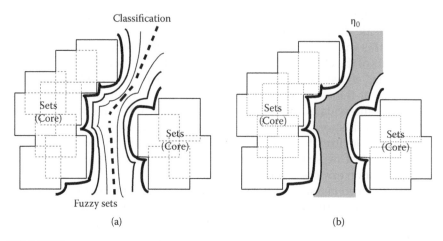

FIGURE 7.19
From sets to fuzzy sets: a principle of a two-level granular classifier exploiting the successive usage of the formalisms of information granulation (a), and further refinements of the information granules realized on the basis of the membership degrees (b).

classification mechanism invoked for any **x** depends on its location, especially whether it is located in the core region or not. Before proceeding with the detailed classification rule, we characterize the two categories of regions with regard to their classification abilities. We start with the quantification of the classification content of information granules (prototypes) and then look at the characterization of discriminatory abilities of the regions located outside the core region. Figure 7.20 offers a detailed view along with all required notation. Let us consider that there are "p" classes in the classification problem under discussion.

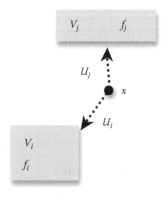

FIGURE 7.20
Core region composed of prototypes and associated notation used in the description of the classification algorithm.

7.7.1 Classification Content of Information Granules

Let us start with associating with the *i*-th granular prototype $\mathbf{V}_i$ (no matter how it was constructed) all data that are included in it. We count the number of patterns belonging to the corresponding classes, forming the ratios of these counts versus the overall number of patterns included in $\mathbf{V}_i$. In this way, we form a vector of frequencies $\mathbf{f}_i = [f_{i1} \ f_{i2} ... \ f_{ip}]$. The quality of the core (granular prototype) depends on the entries of the vector. In an ideal case only a single entry of $\mathbf{f}_i$ is nonzero. The more evident the departure from this situation, the more heterogeneous the prototype is. In the extreme case, we have all equal entries of $\mathbf{f}_i$ the same. The heterogeneity of the prototype can be controlled by varying the size of the granule—the larger the granule, the lower its specificity. Also, in our further considerations, we assume that an overlap between any two granular prototypes $\mathbf{V}_i$ and $\mathbf{V}_j$ does not happen.

7.7.2 Determination of Interval-Valued Class Membership Grades

We compute the interval-valued membership grades in case the data $\mathbf{x}$ is not fully included in the granular prototype. Recall that for the *j*-th variable the bounds of the granular prototype $\mathbf{V}_i$ form the interval $[v_{ij}^-, v_{ij}^+]$ (see Figure 7.20). For the *j*-th coordinate of $\mathbf{x}$, x_j, we consider two situations:

(i) $x_j \notin [v_{ij}^-, v_{ij}^+]$ The bounds of the distance are taken by considering the pessimistic and optimistic scenario and computing the distances from the bounds of the interval, that is, $\min((x_j - v_{ij}^-)^2, (x_j - v_{ij}^+)^2)$ and $\max((x_j - v_{ij}^-)^2, (x_j - v_{ij}^+)^2)$.

(ii) $x_j \in [v_{ij}^-, v_{ij}^+]$ It is intuitive to accept that the distance is equal to zero (as x_j is included in this interval).

The distance being computed on a basis of all variables $||\mathbf{x} - \mathbf{V}_i||^2$ is determined coordinatewise by involving the two situations outlined above. The minimal distance obtained in this way is denoted by $d_{min}(\mathbf{x}, \mathbf{V}_i)$ while the maximal one is denoted by $d_{max}(\mathbf{x}, \mathbf{V}_i)$. More specifically we have (see also Section 10.5 in Chapter 10)

$$d_{min}(\mathbf{x}, \mathbf{V}_i) = \sum_{j \in K} \min((x_j - v_{ij}^-)^2, (x_j - v_{ij}^+)^2)$$

$$d_{max}(\mathbf{x}, \mathbf{V}_i) = \sum_{k \in K} \max((x_j - v_{ij}^-)^2, (x_j - v_{ij}^+)^2) \tag{7.16}$$

where $K = \{j = 1,2,\ldots,n \mid x_j \notin [v_{ij}^-, v_{ij}^+]\}$. Having the distances determined, we compute the two expressions

$$w_1(x) = \frac{1}{\sum_{j=1}^{c} \left(\frac{d_{min}(x,V_i)}{d_{min}(x,V_j)} \right)^{1/(m-1)}}$$

$$w_2(x) = \frac{1}{\sum_{j=1}^{c} \left(\frac{d_{max}(x,V_i)}{d_{max}(x,V_j)} \right)^{1/(m-1)}} \qquad (7.17)$$

Notice that these two formulas resemble the expression used to determine the membership grades in the FCM algorithm. In the same way as in the FCM, the weighted Euclidean distance is considered here, namely, $d_{min}(x, V_i) = \sum_{j\in K} min((x_j - v_{ij}^-)^2/\sigma_j^2, (x_j - v_{ij}^+)^2/\sigma_j^2)$ and $d_{max}(x, V_i) = \sum_{j\in K} max((x_j - v_{ij}^-)^2/\sigma_j^2, (x_j - v_{ij}^+)^2/\sigma_j^2)$ with σ_j being the standard deviation of the j-th variable. These two are used to calculate the lower and upper bounds of the interval-valued membership functions (induced by the granular prototypes). Again, one has to proceed carefully with this construct. Let us start with a situation when x is not included in any of the granular prototypes. In this case, we compute the bounds

$$u_i^-(x) = min(w_1(x), w_2(x))$$

and

$$u_i^+(x) = max(w_1(x), w_2(x)) \qquad (7.18)$$

So, in essence, we arrive at the granular (interval-valued) membership function of the form $U_i(x) = [u_i^-(x), u_i^+(x)]$. If x belongs to V_i then apparently $u_i^-(x) = u_i^+(x) = 1$ (and this comes as a convincing assignment as x is within the bounds of the granular prototype). Obviously, in this case, $u_j^-(x)$ as well as $u_j^+(x)$ for all indexes "j" different from "i" are equal to zero.

7.7.3 Computing Granular Classification Results

Having the prerequisites for the character of the regions of the feature space, we develop the classification method. Two underlying rules are established. The first one concerns x, which is included (covered) by one of the granular prototypes, that is, $x \in X_i$. In this case, the class membership of this data point is of numeric character with the membership grades being the corresponding entries of f_i. In the other case, the calculations of class membership lead to the granular membership grades. First, we determine the bounds

u_i^- and u_i^+, $U_i = [u_i^-, u_i^+]$ for all granular prototypes. Next, we aggregate those with the vectors of class membership $\mathbf{f}_i$ in the following form:

$$\Omega_i = \sum_{i=1}^{c} U_i \bullet \mathbf{f}_i \qquad (7.19)$$

which results in the class membership bounds for each class expressed as

$$\Omega_i = \left[\sum_{i=1}^{c} u_i^- \mathbf{f}_i, \sum_{i=1}^{c} u_i^+ \mathbf{f}_i \right] \qquad (7.20)$$

Obviously some other more advanced aggregation mechanisms of $\mathbf{f}_i$ and U_i could be sought.

In case only some final numeric class membership are of interest, the interval of values shown in Equation (7.20) can be translated into a single numeric value $\tilde{\omega}_i = g(\omega_i^-, \omega_i^+)h(\omega_i^-, \omega_i^+)$ where "g" is a function generating a numeric representative of the interval, that is, $\tilde{\omega}_i = (\omega_i^- + \omega_i^+)/2$ while "h" is a decreasing function of the length of the interval U_i.

It is apparent that the classification scheme developed here delivers very different character classification results depending upon the location $\mathbf{x}$ vis-à-vis the granular prototypes. It is intuitive that while $\mathbf{x}$ is covered by V_i then the results of classification are implied by the classification content of this granular prototype, however when being outside all granular prototypes, the granularity of the class membership (interval-valued membership grades) provide a qualitative as well as quantitative view of the classification results.

7.8 Conclusions

Pattern recognition and its classification schemes function in a highly dimensional feature space. Understanding the geometry of the patterns (learning set) is crucial to further successful design process and a prudent interpretation of classification results. We showed that both in the design and interpretation information granularity plays an essential role. The topology of the patterns with a clear distinction between the regions of high homogeneity and those exhibiting significant heterogeneity becomes helpful. Information granularity is instrumental in the formation of granular prototypes of patterns, which become reflective of the character of the data. Class membership, not only of a numeric nature, but also manifesting a granular character is essential in the quantification of the credibility associated with classification architectures.

References

Abe, S., R. Thawonmas, and Y. Kobayashi. 1998. Feature selection by analyzing class regions approximated by ellipsoids. *IEEE Trans. Systems, Man and Cybernetics-C,* 28, 2, 282–287.

Bargiela, A. and W. Pedrycz. 2003. Recursive information granulation: Aggregation and interpretation issues. *IEEE Trans. Systems, Man and Cybernetics-B,* 33, 1, 96–112.

Bargiela, A. and W. Pedrycz. 2005. A model of granular data: A design problem with the Tchebyschev FCM. *Soft Computing,* 9, 155–163.

Bellman, R., R. Kalaba, and L. Zadeh. 1966. Abstraction and pattern classification. *Journal of Mathematical Analysis and Applications,* 13, 1, 1–7.

Duda, R.O., P.E. Hart, and D.G. Stork. 2001. *Pattern Classification,* 2nd ed. New York: John Wiley.

Fukunaga, K. 1990. *Introduction to Statistical Pattern Recognition,* 2nd ed. New York: Academic Press.

Gabrys, B. and A. Bargiela. 2000. General fuzzy min-max neural networks for clustering and classification. *IEEE Trans. Neural Networks,* 11, 769–783.

Kacprzyk, J. and R.R. Yager. 2001. Linguistic summaries of data using fuzzy logic. *Int. J. General Systems,* 30, 33–154.

Kacprzyk, J. and S. Zadrozny. 2005. Linguistic database summaries and their proto-forms: Toward natural language based knowledge discovery tools. *Information Sciences,* 173, 281–304.

Kandel, A. 1982. *Fuzzy Techniques in Pattern Recognition.* Chichester, UK: Wiley.

Kennedy, J. and R.C. Eberhart. 1995. Particle swarm optimization. *Proc. IEEE Int. Conf. on Neural Networks.* Piscataway, NJ: IEEE Press, Vol. 4, 1942–1948.

MIT-BIH ECG Arrhythmia Database. Beth Israel Hospital, Biomedical Engineering Div. Rm, KB-26. Boston MA.

Pedrycz, W. 1990. Fuzzy sets in pattern recognition: Methodology and methods. *Pattern Recognition,* 23, 1–2, 121–146.

Pedrycz, W. and A. Bargiela. 2002. Granular clustering: A granular signature of data. *IEEE Trans. Systems, Man and Cybernetics,* 32, 212–224.

Pedrycz, W., B.J. Park, and S.K. Oh, 2008. The design of granular classifiers: A study in the synergy of interval calculus and fuzzy sets in pattern recognition. *Pattern Recognition,* 41, 3720–3735.

Pedrycz, W. and G. Succi. 2005. Genetic granular classifiers in modeling software quality. *Journal of Systems and Software,* 76, 277–285.

Rizzi, A., F.M.F. Mascioli, and G. Martinelli. 2000. Generalized min-max classifiers. *Proc. 9th IEEE Int. Conference on Fuzzy Systems, Fuzz-IEEE 2000,* Vol. 1, pp. 36–41.

Roubens, M. 1978. Pattern classification problems and fuzzy sets. *Fuzzy Sets and Systems,* 1, 239–253.

Simpson, P.K. 1992. Fuzzy min-max neural networks-Part 1; Classification. *IEEE Trans. Neural Networks,* 3, 776–786.

Simpson, P.K. 1993. Fuzzy min-max neural networks-Part 2; Clustering. *IEEE Trans. Fuzzy Systems,* 1, 32–45.

Van den Bergh, F. and A.P. Engelbrecht. 2006. A study of particle swarm optimization particle trajectories. *Information Sciences,* 176, 8, 937–971.

8

Granular Models: Architectures and Development

Models are sources of knowledge. Systems are perceived from different points of view and various variables are used in their modeling. In modeling, different levels of specificity of the underlying phenomena are focused on. An ultimate objective is to build a general model, which delivers a holistic view of the system while reconciling some differences and quantifying the still existing diversity of views conveyed by various models. This naturally builds a hierarchical structure where the general model forms the upper level of the hierarchy. Its design invokes various mechanisms of interaction through which the reconciliation between the existing sources of knowledge is realized. The global model has to be of a general nature with an ability not only to reflect the diversity of the existing sources of knowledge but also to quantify this diversity and incorporate it into the global model. To effectively address these challenges and offer a sound design framework, we introduce a concept of *granular* models and show that information granularity is an inherent conceptual and algorithmic component of such models. The principle of justifiable granularity is the underlying development principle. We discuss various scenarios of design of granular models, especially distinguishing between passive and active modes of interaction. In the *passive mode*, the aggregation is accomplished through the use of the principle of justifiable granularity while no feedback mechanism is involved (the original models—sources of knowledge are not modified). In the *active approach* the original sources of knowledge involved become affected and could be modified based on the feedback mechanism already formed at the higher level of the overall hierarchy. Various architectural versions of granular modeling are covered as well.

8.1 The Mechanisms of Collaboration and Associated Architectures

Before proceeding with the detailed conceptual and algorithmic considerations, it is beneficial to make some observations of a general character, which helps motivate the study and focus on the essence of the required design features.

Let us consider a complex system for which is formed a series of models. The system can be perceived from different, perhaps quite diversified points of view. It can be observed over some time periods and analyzed with attention being paid to different levels of detail. Subsequently, the resulting models are built with different objectives in mind. They offer some particular albeit useful views of the system. We are interested in forming a holistic model of the system by taking advantage of the individual sources of knowledge—models, which have been constructed so far. When doing this, we are obviously aware that the sources of knowledge exhibit diversity and hence this diversity has to be taken into consideration and carefully quantified. No matter what the local models look like, it is legitimate to anticipate that the global model (emerging at the higher level of hierarchy) is more general, abstract, with less attention paid to details. From the perspective of the concept of the global model, it is also fair to assume that the model has to engage a sort of formalism of information granules to represent and in this way quantify the diversity of the views of the system.

The overall process comes hand in hand with some mechanisms of collaboration, reconciliation of individual sources of knowledge, or knowledge transfer.

This brings up the concepts of *granular* fuzzy models—a new modeling paradigm in fuzzy modeling in which we form a hierarchy of modeling constructs starting with fuzzy models at the lower level of the hierarchy and ending up with a granular fuzzy model emerging at the higher level.

We identify three general models of collaboration and knowledge transfer, namely,

(a) *Realization of granular models at the higher level of hierarchy.* The individual models at the lower level could be quite diversified. Depending upon the nature of the model at the higher level, we talk about *granular* fuzzy models, *granular* regression, *granular* networks, and so forth.

(b) *Realization of a granular model at the same level as the original local models.* Information granularity of the granular model reflects not only the diversity of the sources of knowledge (models) but also facilitates active involvement and collaboration among individual models. Some examples in this class of models include granular Analytic Hierarchy Process (AHP) decision-making models and granular fuzzy cognitive maps.

(c) *Realization of granular models being a direct result of knowledge transfer.* Here the component of information granularity becomes a result of the formation of a model whose outcomes become reflective of a limited confidence in the model given the new scenario though it is somewhat similar to the previous one.

While the first class of granular models will be discussed in detail in this chapter, the two other schemes will be presented in depth in the successive chapters.

In the hierarchical reconciliation of knowledge, we distinguish two general approaches, which depend on the way in which the knowledge is being utilized. The corresponding visualization of the essence of this mechanism is presented in Figure 8.1.

Passive approach. In this approach, Figure 8.1a, we are provided with the sources of knowledge, which are used effectively at the higher level of the hierarchy producing a certain granular model. The models at the lower level are not affected so the resulting granular model does not entail any possible modifications, updates, or refinements of the original sources of knowledge located at the lower level of hierarchy.

Active approach. In contrast to the passive approach, here we allow for some mechanisms of interaction. The results of modeling realized at the higher level of the hierarchy are used to assess the quality of the local models present at the lower level. The feedback (visualized by a dotted line in Figure 8.1b) is delivered to the lower level where the local knowledge (models) present at the lower level becomes affected and modified.

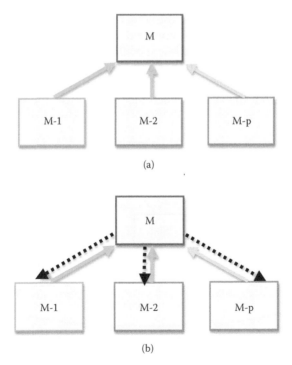

FIGURE 8.1
Passive (a) and active (b) hierarchical knowledge reconciliation through interaction among models.

8.2 Realization of Granular Models in a Hierarchical Modeling Topology

As already highlighted, this component of information granularity delivers a desired level of flexibility required to reflect and quantify an inherent diversity of individual models we have to deal with. The underlying architectural layout is visualized in Figure 8.2. The unavoidable information granularity effect manifests at the resulting model formed at the higher level of abstraction, hence the construct obtained there comes as a generalization of models present at the lower level of the hierarchy, namely, *granular* models.

A certain system or phenomenon is perceived from different points of view (perspectives). Subsequently, the models emerging there and denoted as M-1, M-2, ..., M-p use locally available data D-1, D-2, ..., D-p. In general, we can envision these models to be of different nature (i.e., rule-based architectures, neurofuzzy models, fuzzy cognitive maps, regression models). The models M-1, M-2, ..., M-p are now brought together (reconciled) by forming a global model at the higher level of hierarchy. Not engaging in more specific discussion, some general and intuitively convincing observations could be made. The model at the higher level is definitely more abstract (general) than those at the lower level as it tries to embrace and quantify a variety of sources of knowledge (models). As such we envision that the output of such model is less specific than the numeric output generated by any of the models positioned at the lower level. This is not surprising at all. Imagine that the input is equal to **x** with **x** being a numeric vector of inputs coming from the system. Each model produces the numeric outputs

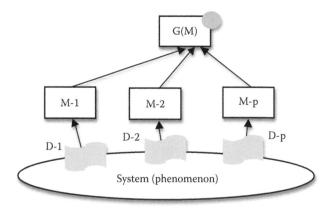

FIGURE 8.2

The emergence of information granularity at the model G(M) at the upper level of the hierarchy coming as a result of dealing with the diversity of models M-1, M-2, ..., M-p. The emergence of information granularity is denoted by a large gray dot placed next to the granular model being formed there.

M-1(**x**), M-2(**x**), ..., M-p(**x**). In virtue of the different perception of the system itself, it is very likely that these inputs are different; we do not rule out that they could be close to each other. If the model at the higher level is to capture this diversity of the sources of knowledge, intuitively for the same **x** it should return an output that is more abstract than a single numeric entity, that is, a certain information granule. Hence, such models producing granular outputs will be referred to as *granular* models.

We contrast the proposed construct with the approaches existing in the literature, because, on the surface, there seem to be some similarities (Alcala et al., 2009; Pham and Castellani, 2006; Roubos and Setnes, 2001). There are several evident and well-delineated differences, though:

- The result of aggregation is positioned at the higher level of abstraction in comparison with the results generated by the individual models. For instance, the models involved in the aggregation produce numeric outputs and numeric outputs are also generated as a result of aggregation. Here, the outputs are granular (intervals, fuzzy sets, rough sets, etc.) even though the individual models generate plain numeric outputs. This can be viewed as a substantial advantage of the overall topology as it helps quantify the diversity of the individual sources of knowledge. The quantification of this diversity is naturally realized by invoking the principle of justifiable granularity (as it will be discussed in detail in the next section).

- The data are not shared meaning that each model could be constructed on the basis of locally available data; furthermore, the variables at each data set need not be the same. This stands in sharp contrast with the scenario of bagging or boosting a family of models.

- The individual models could be different (i.e., neural networks, rule-based models, linear regression, etc.).

8.3 The Detailed Considerations: From Fuzzy Rule-Based Models to Granular Fuzzy Models

Proceeding with more detailed investigations based on the general scheme discussed in the previous section, we present one of the interesting scenarios for capturing sources of knowledge (fuzzy models) in the case when the individual sources of knowledge-fuzzy models are treated as a collection of fuzzy rule-based models (Pedrycz and Rai, 2008). The management of knowledge realized here is focused on the use of information granules forming a backbone of the individual fuzzy models at the lower level. As the

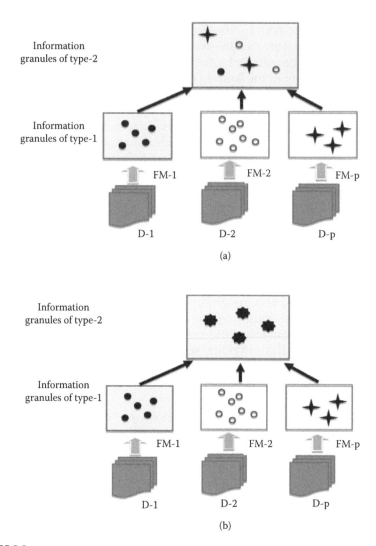

FIGURE 8.3
Formation of the information granules at the higher level: (a) selection and (b) clustering of prototypes.

antecedents of the rules are formed on the basis of information granules, the realization of a certain model at the higher level of hierarchy calls for the formation of a collection of information granules to start with. Here, we envision two main directions in the construction of information granules as portrayed in Figure 8.3.

A selection of a suitable subset of information granules forms the individual models, Figure 8.3a. The prototypes of the information granules are selected in such a way that they represent all prototypes of the models to the highest extent. This is a combinatorial optimization problem, which may call for techniques of evolutionary optimization (e.g., genetic algorithms, GAs) or population-based optimization (e.g., particle swarm optimization, PSO) to arrive at solutions to the problem. The optimization criterion quantifies a reconstruction error of the prototypes at the lower level when being expressed in terms of the subset of the prototypes formed at the upper level.

The second approach, illustrated in Figure 8.3b, is concerned with clustering (granulation) of prototypes available at the lower level. The standard FCM can be used here. The method operates on a family of the prototypes present in all fuzzy models positioned at the lower level of the hierarchy and produces "c" prototypes at the upper level of the hierarchy. In light of the construction of the information granules, which have been built at the higher level (which are also sought as a more abstract view of the information granules present at the lower level), we may refer to them as information granules of higher type, that is, type-2 information granules (and type-2 fuzzy sets, in particular).

Given the collection of information granules (which can be represented as a family of the prototypes), we are in a position to develop a model at the higher level. We note there is an inherent granularity of the associated model, which comes from the fact that for any prototype formed at the higher level $\mathbf{v}_i$, each fuzzy model at the lower level returns some numeric value, that is, $M\text{-}1(\mathbf{v}_1)$, $M\text{-}2(\mathbf{v}_2)$,…, $M\text{-}p(\mathbf{v}_i)$. It is very unlikely that all these values are the same. This set of data is subject to the granulation process (with G denoting a granulation mechanism used in the realization of the principle of justifiable granularity). The size (level of specificity) of the resulting information granule depends upon the predetermined value of the parameter α, which was used to construct information granules. We will be taking advantage of this flexibility in the realization of the model guided by two conflicting objectives of coverage offered by the model and the specificity of the results produced there.

Overall, for "c" information granules built at the higher level, the available experimental evidence (originating at the lower level of hierarchy) arises in the form

$$\{(\mathbf{v}_1, G\{FM\text{-}1(\mathbf{v}_1), FM\text{-}2(\mathbf{v}_1),..,FM\text{-}p(\mathbf{v}_1)\}),$$
$$(\mathbf{v}_2, G\{FM\text{-}1(\mathbf{v}_2), FM\text{-}2(\mathbf{v}_2), …, FM\text{-}p(\mathbf{v}_2)\}), …$$
$$(\mathbf{v}_i, G\{FM\text{-}1(\mathbf{v}_i), FM\text{-}2(\mathbf{v}_i), …, FM\text{-}p(\mathbf{vi}_1)\}), …$$
$$(\mathbf{v}_c, G\{FM\text{-}1(\mathbf{v}_c), FM\text{-}2(\mathbf{v}_c), …, FM\text{-}p(\mathbf{v}_c)\})\} \qquad (8.1)$$

and invokes an evident component of granularity as highlighted in Figure 8.4. Taking these data into consideration, we construct a granular model. Preferably, the model of this nature has to be structure-free as much as possible. In this case, a technique of case-based reasoning (CBR), or being more specific, granular case-based reasoning can be contemplated

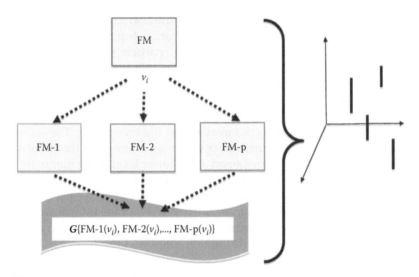

FIGURE 8.4
Experimental evidence behind the formation of the granular model to be constructed at the higher level. Considered here is an interval format of resulting information granules.

(Figure 8.5). The other option worth considering is the concept of fuzzy regression. Moving to the computational details, we compute a degree of activation of the cases by a certain input **x**.

$$u_i(\mathbf{x}) = \frac{1}{\sum_{j=1}^{c} \left(\frac{\|\mathbf{x}-\mathbf{v}_i\|}{\|\mathbf{x}-\mathbf{v}_j\|} \right)^{2/(m-1)}}, \quad m>1 \tag{8.2}$$

As the outputs are evidently granular (intervals), we determine the lower and upper bound of the granular model based on the bounds of y_i^- and y_i^+

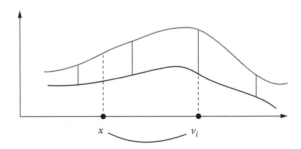

FIGURE 8.5
The mechanism of granular case-based reasoning and a construction of lower and upper bounds.

where $[y_i^-, y_i^+] = G\{FM\text{-}1(\mathbf{v}_i), FM\text{-}2(\mathbf{v}_i),..,FM\text{-}p(\mathbf{v}_i)\}$. We obtain the following expressions for the bounds,

$$\text{lower bound: } y^-(\mathbf{x}) = \sum_{i=1}^{c} u_i(\mathbf{x})y_i^- \qquad (8.3)$$

$$\text{upper bound: } y^+(\mathbf{x}) = \sum_{i=1}^{c} u_i(\mathbf{x})y_i^+ \qquad (8.4)$$

An example of the granular *envelope* produced in this way is shown in Figure 8.6. Note that the parameter "m" (fuzzification coefficient used in the FCM method) affects the shape of the granular mapping (and this could be used as a source of additional parametric flexibility built into the granular model).

The granularity of the data themselves depends upon the value of α being used when running the process of justifiable granularity. In this modeling scenario, we are faced with a two-objective optimization problem where one of the objectives is to make the envelope as narrow as possible (to achieve high specificity level, which is desirable) and at the same time *cover* as much experimental evidence as possible (so a large number of data are included within the lower and upper bound making the granular model highly legitimized by the experimental evidence captured by the series of fuzzy models at the lower level of the hierarchy). Intuitively, the two requirements, which are evidently in conflict, are illustrated in Figure 8.7.

The satisfaction of the two requirements and the way of achieving a certain compromise can be controlled by choosing a certain value of α used in the realization of the principle of the justifiable granularity. To proceed with a numeric quantification of these two criteria, we use (a) the cumulative length of information granules (intervals) of the output of the granular fuzzy model $L = \sum_i L_i$, and (b) the coverage level of the data taken as a ratio I, $I = \frac{\text{no. of data covered}}{\text{all data}}$. The higher this ratio, the better the coverage of the data. Ideally, we may like to see I be very close to 1, however, this might result in an unacceptable lack of specificity of the results provided by the granular fuzzy model. Accepting a certain value of α in the construction of the information granules through running the principle of justifiable granularity, we calculate the resulting values of I as well as the cumulative length of intervals of the granular outputs of the model. The calculations of the area under the curve (AUC) lead to the global quantification of the granular model: we say that model F is better than F′ if AUC(F) > AUC(F′).

As an example illustrating the construction of the granular fuzzy mode, we consider a concrete compressive strength data (http://archive.ics.uci.edu/ml/datasets/Concrete+Compressive+Strength). This data set coming from the Machine Learning Repository concerns a compressive strength of concrete in association with its characteristics such as blast furnace slag,

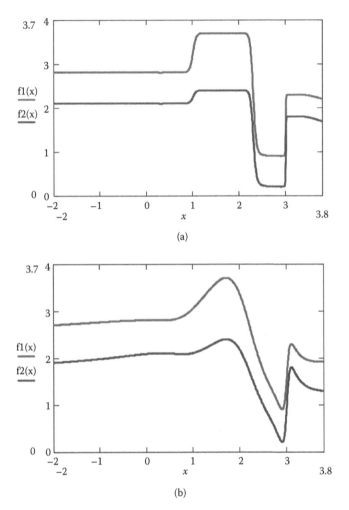

FIGURE 8.6
Plots of the granular model for selected values of m: m = 1.2 (a) and 2.0 (b).

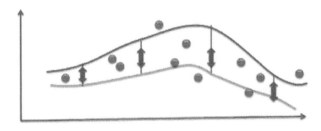

FIGURE 8.7
Illustration of the two objectives (coverage and specificity) to be optimized in the construction of the granular model.

fly ash, water, superplasticizer, coarse aggregate, and fine aggregate. It consists of 1,030 pairs of input–output data. The FCM algorithm was used in the experiments; the fuzzification coefficient was set to 2.0. We form 10 fuzzy rule-based models where each of them is constructed on the basis of 103 randomly selected data. The number of rules in each model is equal to the number of fuzzy clusters determined for each locally available data set. The performance index Q, which is used to evaluate the quality of the model is expressed in the following form:

$$Q = \frac{1}{N}\sum_{k=1}^{N}(\hat{y}_k - y_k)^2 \qquad (8.5)$$

where N is the number of data present in the individual data set. Once the information granules have been formed by using the fuzzy c-means (FCM) clustering algorithm, the parameters of the linear functions present in the conclusion parts of the rules were estimated by running a standard least square error (LSE) method. The number of rules itself is subject to optimization and here we resort to a successive enumeration by designing the fuzzy model for increasing the number of rules while monitoring the values of the corresponding performance index Equation (8.5). The values of Q treated as a function of the number of rules (clusters) for all the models are shown in Figure 8.8.

One can note (as could be expected) that when the number of clusters increases, the values of the performance index decrease. Essentially all the local models follow the same tendency: there is a cutoff point at c = 5 or 6 beyond which the values of the performance index are not significantly affected by further increasing the number of rules. To visualize the performance of the fuzzy models formed here, a collection of plots of output data versus the corresponding outputs of the models is included in Figure 8.9.

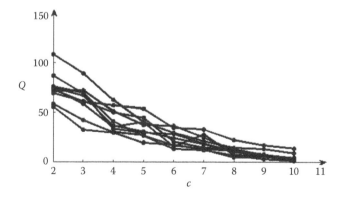

FIGURE 8.8

The performance index of the models versus the number of rules (clusters). Shown are the values for all 10 models.

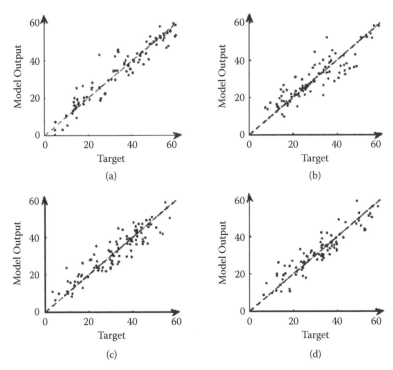

FIGURE 8.9
The performance of selected fuzzy models (data versus model output): (a) Q = 21.94, (b) Q = 31.16, (c) Q = 29.25, (d) Q = 24.01.

Once the 10 local models have been constructed, the prototypes present there are clustered at the higher level to form a backbone of the granular fuzzy model. Following the principle of justifiable granularity, for each of the prototypes obtained in this way we form the intervals of the corresponding output values.

The characterization of the quality of the overall model can be conveyed by plotting the values of the coverage of data I versus the cumulative length of the intervals produced by the granular fuzzy model for the inputs formed as the prototypes of the information granules obtained for the individual fuzzy models. Some selected relationships of this form are shown in Figure 8.10.

8.4 A Single-Level Knowledge Reconciliation: Mechanisms of Collaboration

The collaboration schemes presented so far are of hierarchical character. No matter whether a passive or active mode was considered, the granular model is built at the higher level of the hierarchy. In contrast to the hierarchical

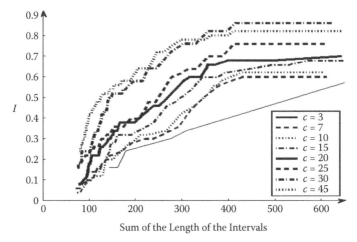

FIGURE 8.10
Coverage of data I versus the cumulative length of intervals L of the output for a selected number of clusters.

mode of collaboration (discussed in Section 8.1 in Chapter 8), the mechanisms presented here can be referred to as a one-level collaboration, Figure 8.11. To focus our presentation, discussed are fuzzy rule-based models (sources of knowledge). Here the collaboration in the formation of fuzzy models is mostly focused on the collaborative formation of information granules as they form a backbone of fuzzy models. The conclusion parts are mostly realized based on the locally available data and they are constructed once the information granules have been established.

There are a number of possible mechanisms of interaction between the individual models when exchanging the findings about the structure of information granules. The form of interaction depends on the level of compatibility

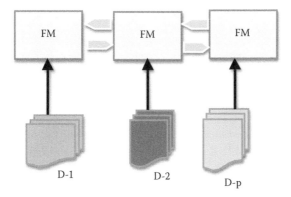

FIGURE 8.11
Collaboration between fuzzy models realized through interaction performed at the level of information granules.

considering available spaces of data and spaces of features (inputs) and commonalities among them.

The findings of the corresponding character are exchanged (communicated among the sources of knowledge) and actively used when carrying out information granulation at the level of the individually available data sets. In what follows, we elaborate on the key aspects of the collaboration modes referring the reader to the literature on their algorithmic details (Pedrycz and Rai, 2008). The taxonomy provided here is based on the commonalities encountered throughout the individual data sources. The two main modes are distinguished. They could be present in terms of the same feature space or the same data being expressed in different feature spaces.

Collaboration through exchange of prototypes. As presented in Figure 8.12, the data are described in the same feature space and an interaction is

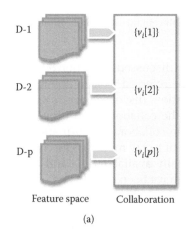

(a)

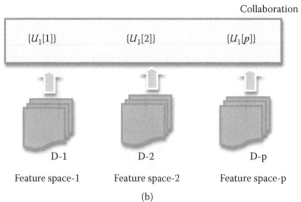

(b)

FIGURE 8.12
A schematic view of collaboration through (a) exchange of prototypes and (b) partition matrices.

realized through communicating the prototypes, which were produced locally.

Collaboration through exchange of partition matrices. Here the data are described in different feature spaces (they might overlap but are not the same). The data in each data set are the same but described in different feature spaces. The exchange of findings and collaboration is realized through interaction at the level of partition matrices. Note that these matrices abstract from the feature spaces (the spaces do not appear there in an explicit way) but the corresponding rows of the partition matrices have to coincide (meaning that we are concerned with the same data).

Formally, the underlying optimization problem can be expressed by an augmented objective function, which is composed of two components,

$$Q = Q(D - ii) + \beta \sum_{\substack{jj=1, \\ jj \neq ii}}^{p} \| G(ii) - G(jj) \|^2 \tag{8.6}$$

The first one, $Q(\mathbf{D}\text{-ii})$ is focused on the optimization of the structure based on the locally available data (so the structure one is looking at is based on $\mathbf{D}$-ii). The second one is concerned with achieving consistency between the granular structure $\mathbf{G}(ii)$ and the structure revealed based on other data. The positive weight (β) is used to set up a certain balance between these two components of the augmented objective function (local structure and consistency among the local structures). The notation $\mathbf{G}(ii)$ is used to concisely denote a collection of information granules obtained there, that is, $\mathbf{G}(ii) = \{G_1[ii], G_2[ii],..., G_c[ii]\}$. As mentioned, such granules could be represented by their prototypes or partition matrices.

Proceeding with the FCM-like optimization, the objective function can be written in a more explicit fashion as follows:

$$Q = \sum_{\substack{i=1 \\ }}^{c} \sum_{\substack{k=1 \\ x_k \in D-ii}}^{N} u_{ik}^m \| x_k - v_i[ii] \|^2 + \beta \sum_{\substack{jj=1, \\ jj \neq ii}}^{p} \| G(ii) - G(jj) \|^2 \tag{8.7}$$

In case of communication at the level of the prototypes, Figure 8.12a, the objective function becomes refined and its term guiding the collaboration effect arises in the form

$$\sum_{\substack{jj=1, \\ jj \neq ii}}^{p} \| G(ii) - G(jj) \|^2 = \sum_{i=1}^{c} \sum_{\substack{jj=1, \\ jj \neq ii}}^{p} \| v_i[ii] - v_i[jj] \|^2 \tag{8.8}$$

For the communication with the aid of partition matrices, Figure 8.12b, the detailed expression for the objective function reads as follows:

$$\sum_{\substack{jj=1,\\ jj\neq ii}}^{P} \| G(ii) - G(jj) \|^2 = \sum_{i=1}^{c} \sum_{k=1}^{N} \sum_{\substack{jj=1,\\ jj\neq ii}}^{P} (u_{ik}[ii] - u_{ik}[jj])^2 \| \mathbf{v}_i[ii] - \mathbf{v}_j[jj] \|^2 \qquad (8.9)$$

It could be noted that there is a certain direct correspondence between the prototypes and the partition matrix in the sense that each one could be inferred given that the other one has been provided. More specifically, we envision the following pair of mappings supplying equivalence transformations (see also Figure 8.13):

$$\{U, D\} \rightarrow V = \{\mathbf{v}_1, \mathbf{v}_2, ..., \mathbf{v}_c\} \qquad \{V, D\} \rightarrow U \qquad (8.10)$$

This transformation can bring a certain unified perspective to the mechanisms of exchange of information granules. For instance, one can convey a collection of the prototypes and they can induce a partition matrix over any data set.

An interesting collaboration scenario in which prototypes are a communication vehicle leads to a concept of clustering with viewpoints (Pedrycz, Loia, and Senatore, 2010). In the optimization problem resulting here, we incorporate the external structural knowledge directly into the minimized objective function.

Consider an illustrative example shown in Figure 8.14. The structure to be built based on locally available data D is influenced by prototypes—viewpoints coming from the structure constructed at some other data set.

By considering the viewpoints to be of the same relevance as the prototypes themselves, we position them on the list of the prototypes. From a formal standpoint, the viewpoints are conveniently captured by two matrices; let us denote these matrices by B and F. The first one is of Boolean nature and comes in the following form:

$$b_{ij} = \begin{cases} 1, & \text{if the j-th feature of the i-th row of B is determined by the viewpoint} \\ 0, & \text{otherwise} \end{cases}$$

$$(8.11)$$

FIGURE 8.13
Mappings between components of granular representations of data.

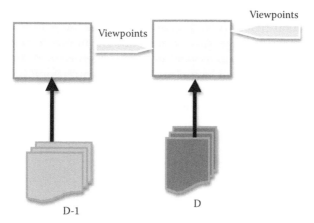

FIGURE 8.14
Example of conveying local structural funding with the use of viewpoints—prototypes provided by some other models built on local data.

Thus, the entry of B equal to 1 indicates that the (i, j)-th position is determined by a certain viewpoint. The dimensionality of B is $c \times n$. The second matrix (F) is of the same dimensionality as B and includes the specific numeric values of the available viewpoints. For instance, consider a collection of six-dimensional data. If $c = 5$ and there are three viewpoints defined only for the second variable and having the values a_1, a_2, and a_3, respectively, then this structural information is conveyed in the following way:

$$
B = \begin{bmatrix}
0 & 1 & 0 & 0 & 0 & 0 \\
0 & 1 & 0 & 0 & 0 & 0 \\
0 & 1 & 0 & 0 & 0 & 0 \\
0 & 0 & 0 & 0 & 0 & 0 \\
0 & 0 & 0 & 0 & 0 & 0
\end{bmatrix}
\quad
F = \begin{bmatrix}
0 & a_1 & 0 & 0 & 0 & 0 \\
0 & a_2 & 0 & 0 & 0 & 0 \\
0 & a_3 & 0 & 0 & 0 & 0 \\
0 & 0 & 0 & 0 & 0 & 0 \\
0 & 0 & 0 & 0 & 0 & 0
\end{bmatrix}
$$

If we concentrate on the fuzzy c-means clustering, the corresponding objective function is formed in a way that it captures the viewpoints, which are formally expressed through the viewpoint-based matrices B and F,

$$
Q = \sum_{k=1}^{N} \sum_{i=1}^{c} \sum_{\substack{j=1 \\ i,j:b_{ij}=0}}^{n} u_{ik}^{f}(x_{kj} - v_{ij})^2 + \sum_{k=1}^{N} \sum_{i=1}^{c} \sum_{\substack{j=1 \\ i,j:b_{ij}=1}}^{n} u_{ik}^{f}(x_{kj} - f_{ij})^2
\qquad (8.12)
$$

There are two parts of the objective function. The first is concerned with the minimization that is directly associated with the data to be clustered (so here the entries of B are equal to zero) and the minimization invokes the prototypes. The second component of Q is concerned with the dispersion between the data and the viewpoints (with the entries of B being equal 1). We rewrite Equation (8.12) in a more compact format by introducing a new matrix G (where $\dim(G) = c \times n$) with the following entries,

$$g_{ij} = \begin{cases} v_{ij} & \text{if } b_{ij} = 0 \\ f_{ij} & \text{if } b_{ij} = 1 \end{cases} \tag{8.13}$$

This leads to the expression

$$Q = \sum_{k=1}^{N} \sum_{i=1}^{c} \sum_{j=1}^{n} u_{ik}^{f}(x_{kj} - g_{ij})^{2} \tag{8.14}$$

As usual, the optimization of Equation (8.14) involves the two components describing information granules, namely, the partition matrix and the prototypes.

8.5 Collaboration Scheme: Information Granules as Sources of Knowledge and a Development of Information Granules of a Higher Type

In the discussion presented in the previous section, we assumed that the number of information granules was the same for all parties involved in the collaboration process. While this assumption might be legitimate under some circumstances, there are situations when there are evidently differences in the levels of information granularity when perceiving a certain system (data). The investigations made so far require some augmentation.

Here we look at one of the possibilities, Figure 8.15, where for the same data set **D** formed are various granular views manifesting in collections of information granules {U[ii]}, ii = 1, 2, ..., p.

Any interaction or a creation of a general view of the individual information granules can be done once we establish a certain conceptual level at which the objects such as proximity matrices can be looked at in a unified manner. Note that, in general, partition matrices are of different dimensionality so that no direct comparison (matching) can be realized. A more general view of the matrices is formed by constructing proximity matrices

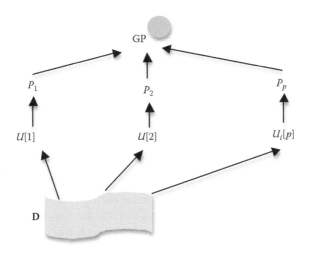

FIGURE 8.15
Collections of information granules formed for data **D** and a realization of a granular abstraction GP through the use of proximity matrices P_1, P_2,..., P_p.

where all of them are of dimensionality NxN. Let us recall that for a given partition matrix U[ii], the induced proximity matrix P[ii] = $[p_{k,l}[ii]]$, k, l = 1, 2, ..., N is defined as

$$p_{k,l}[ii] = \sum_{i=1}^{c_{ii}} \min(u_{ik}[ii], u_{il}[ii]) \tag{8.15}$$

The proximity matrices satisfy the property of symmetry $p_{k,l}[ii] = p_{l,k}[ii]$ and reflexivity $p_{k,k}[ii] = 1$.

As the proximity matrices have the same dimensionality, a global view can be established. The proximity matrices exhibit diversity, which is reflective of the diversity encountered for the partition matrices. The use of the principle of justifiable granularity applied to the corresponding individual entries of the proximity matrices gives rise to a granular proximity matrix GP = G{P[ii], ii = 1, 2,...,p}. Recall that the result depends upon the value of α and GP quantifies the diversity of the structural findings produced at the lower level.

Based on the global view delivered by GP, one can realize some adjustments to the already constructed fuzzy partition matrices to make them more in line with the global view and eventually eliminate some significant differences. This is accomplished by minimizing the following performance index:

$$Q = \sum_{ii=1}^{p} incl(Prox(U[ii]), GP^+) incl(GP^-, Prox(U[ii])) \tag{8.16}$$

We strive to maximize the inclusion of proximity of U[ii], Prox(U[ii]) in the granular construct GP by adjusting the entries of U[ii]. The inclusion operation, incl(,.) standing in the performance index is implemented depending on the nature of the information granule GP. In the simplest case, the inclusion returns 1 once the value of P[ii] is contained in the interval values of GP, $GP = [GP^-, GP^+]$.

This scheme realizes an active mode of interaction as the updated partition matrices can be used in the formation of the granular construct GP and subsequently this new granular proximity matrix is used in the minimization of Equation (8.16), thus leading to the new partition matrices. The dynamics of the process depends on the value of the parameter α used in the realization of the principle of justifiable granularity. Higher values of α impose more requirements on the convergence process involving successive minimization of Equation (8.16). Apparently, $\alpha = 0$ does not invoke any iterative scheme.

In general, the collaboration architecture can be formed for a collection of models as illustrated in Figure 8.16.

Once a probing element x_k coming from $\mathbf{D}$ (= {x_1, x_2, ..., x_N}) is provided to all sources of knowledge (M_1, M_2,..., M_p), then their corresponding output (y_{k1}, y_{k2},..., y_{kp}) are available, that is, $y_{ki} = M(x_{ki})$. For the *i*-th source, we form an information granule Y_{ki} on a basis of (y_{k1}, y_{k2},...,$y_{k/i-1}$, $y_{k/i+1}$,..., y_{kp}) and adjust the parameters of M_i so that the inclusion relationship of y_{ki} in Y_{ki}, incl(y_{ki}, Y_{ki}), is satisfied. We require that the inclusion be satisfied for as many of x_ks as possible, which can be achieved by minimizing the following sum:

$$Q_i = \sum_{k=1}^{N} \text{incl}(y_{ki}, Y_{ki}) \tag{8.17}$$

The optimization is realized for all knowledge sources and the process is repeated. Its convergence is affected by the values of α impacting the specificity of information granules; its higher values impose more vigorous

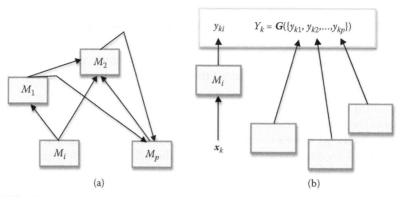

(a)   (b)

FIGURE 8.16
Collaboration among the models guided by information granules: (a) a general scheme, (b) details concerning the optimization of the *i*-th knowledge source.

interaction among information granules. At each iteration, an overall performance index $Q = \sum_{i=1}^{P} Q_i$ is computed in order to assess the effectiveness of the collaboration process.

8.6 Structure-Free Granular Models

Here we introduce and develop a concept of mappings realized in the presence of information granules and offer a design framework supporting the formation of such mappings. Information granules are conceptually meaningful entities (Zadeh, 1997) formed on the basis of a large number of experimental input–output numeric data available for the construction of the model, on the basis of which we form a mapping. Considering the directional nature of the mapping to be formed, the directionality aspect of the mapping needs to be taken into account when developing information granules meaning that while the information granules in the input space could be constructed with a great deal of flexibility, the information granules formed in the output space have to inherently relate to those built in the input space. The input space is granulated with the aid of the FCM clustering algorithm.

The information granules in the output space are constructed with the aid of the principle of justifiable granularity (being one of the underlying fundamental conceptual pursuits of Granular Computing). The construct exhibits two important features. First, the constructed information granules are formed in the presence of information granules already constructed in the input space (and this realization is reflective of the direction of the mapping from the input to the output space). Second, the principle of justifiable granularity does not confine the realization of information granules to a single formalism such as fuzzy sets, for example, but helps form the granules expressed for any required formalism of information granulation. The quality of the granular mapping (namely, the mapping realized for the information granules formed in the input and output spaces) is expressed in terms of the coverage criterion (articulating how well the experimental data are covered by information granules produced by the granular mapping for any input experimental data). Some parametric studies are reported by quantifying the performance of the granular mapping (expressed in terms of the coverage and specificity criteria) versus the values of certain parameters utilized in the construction of output information granules through the principle of justifiable granularity. The plots of coverage—specificity dependency help determine a knee point and reach a sound compromise between these two conflicting requirements imposed on the quality of the granular mapping. Furthermore, the quality of the mapping is quantified with regard to the number of information granules (granularity of the mapping). A series of experiments is reported as well.

The models are expressed in the form of mapping from information granules existing in the input space to the granules present in some output space. There are several tangible advantages of this category of mappings. First, they are general and avoid focusing on details, which come with premature design decisions as to the class of models. Second, the idea of granular mapping applies equally well to different formalisms of information granules. Third, carefully developed performance indexes help quantify the quality of the mapping. Fourth, a level of detail could be adjusted by choosing the number of information granules. Once the granular mapping has been formed, it could be a subject of further refinements leading eventually to more detailed numeric constructs.

8.7 The Essence of Mappings between Input and Output Information Granules and the Underlying Processing

We concentrate on the concepts of information granules used in the realization of a certain granular input–output mapping (relationship) $y = f(x)$ based on some available experimental data present in the multivariable space, namely, (x_k, y_k), $k = 1, 2, \ldots, N$ where $x_k \in \mathbf{R}^n$ and $y_k \in \mathbf{R}$.

The mapping is realized at the level of information granules, which are the essential building blocks of this construct. A view of the granular mapping emphasizing the way it is formed upon the individual information granules constructed in the input and output space is presented in Figure 8.17.

The information granules in the input and output spaces are constructed in different ways. While in the case of the input space, the granules are formed in a fairly flexible manner (as they concern the independent variables), the granules in the output space are developed in a certain way that are reflective of the nature of the dependent variable (and experimental data therein)

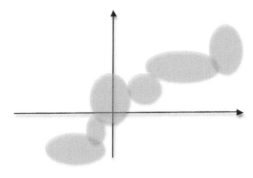

FIGURE 8.17
Example of granular mapping (one-dimensional input space): a general concept.

and its linkage with the input variables or, to be more specific, in a way in which the input space is perceived through information granules.

(a) The input space (where the data x_k are located) is granulated with the use of FCM resulting in "c" fuzzy sets described by the corresponding rows of the partition matrix U, u_i. Each cluster is characterized by (associated with) the distribution of data of the output variable associated with the corresponding input information granule. As a matter of fact, any clustering (Boolean or fuzzy) could be considered here.

(b) As far as the information granules formed in the output space are concerned, that is, B_1, B_2, ..., B_c, their construction has to be linked with the already established granules in the input space (which indirectly links the input and output data). To emphasize this observation, we describe B_i as follows

$$B_i = G(\{y_1, y_2, ..., y_N\}, A_1, A_2, ..., A_c) \tag{8.18}$$

where in the above expression we highlight the elements involved in the resulting construct involved (namely, the data and the information granules in the input space). The symbol G denotes a process of granulation of the data.

Once the information granules have been developed, a mapping from $\mathbf{R}^n$ to $\mathbf{R}$ realized in terms of the input–output information granules can be written down as $Y = F(A_1, A_2, ..., A_c, B_1, B_2,..., B_c)$. For any numeric input $\mathbf{x}$, the result of the mapping F (given the fact that it invokes information granules present there) is also a certain information granule Y. A general category of the mappings (F) is expressed as a *sum* of the corresponding information granules *weighted* by the matching degrees obtained for A_i and $\mathbf{x}$, denoted here as $A_i(\mathbf{x})$, that is,

$$Y = \sum_{i=1}^{c} A_i(\mathbf{x}) \otimes B_i \tag{8.19}$$

The semantics of the operations standing in Equation (8.19) has to be carefully articulated. The aggregation operator $\otimes$ has to be specified once a certain formalism of information granules has been specified. Furthermore, at this stage, the summation operator (Σ) needs to be clarified as well.

8.8 The Design of Information Granules in the Output Space and the Realization of the Aggregation Process

Information granules formed in the output space are inherently associated with the information granules already constructed in the input space and such linkages have to be taken into consideration in the development of the granules. Several interesting main alternatives as far as the formal setup of the information granulation concerned is contemplated:

(a) *Interval form of information granules based on the use of the principle of justifiable granularity.* As a result of the use of the principle (discussed in the following section), we obtain a collection of associated intervals in the output space, namely, $[y_1^-, y_1^+], [y_2^-, y_2^+],...,[y_c^-, y_c^+]$. Following the aggregation scheme Equation (8.19), for any $\mathbf{x}$, we obtain the interval with the bounds expressed as follows, $[y^-, y^+] = [\sum_{i=1}^{c} A_i(\mathbf{x})b_i^-, \sum_{i=1}^{c} A_i(\mathbf{x})b_i^+]$.

(b) *Information granules in the form of fuzzy sets with some membership functions.* The algebraic operations standing in Equation (8.19) are realized in terms of fuzzy arithmetic (Klir and Yuan, 1995) producing a fuzzy set Y with the membership function coming as a result of the use of the extension principle (Zadeh, 1975),

$$Y(y) = \sup_{z_1, z_2,...,z_c \in R} [\min(A_i(\mathbf{x}), B_1(z_1), B_2(z_2),...,B_c(z_c))]$$

subject to constraints

$$y = \sum_{i=1}^{c} A_i(\mathbf{x})z_i \tag{8.20}$$

The computations identified above (maximization under constraints) could be simplified if the fuzzy numbers come in some form (i.e., triangular fuzzy numbers) or one form, a solution by solving the problem for several values of α-cuts of the fuzzy sets standing in the original formula.

(c) *Probabilistic information granules.* Based on the membership degrees $A_1, A_2,..., A_c$ the corresponding fuzzy histograms are constructed $p_1(y), p_2(y),..., p_c(y)$. For the i-th input information granule (fuzzy cluster), we form a fuzzy histogram. In comparison with the methods of building *standard* histograms, here we collect y_ks weighted by the membership grades u_{ik} and collect the results into bins. Using them the probability function $p_i(y)$ is formed. Obviously, a variety of methods used in the estimation of probability density functions (Hastie et al., 2001) both parametric and nonparametric, could be advantageous here.

The determination of the probabilistic granules is realized by forming a mixture of the probability functions weighted by the activation levels of the corresponding input information granules, namely,

$$p(y) = \sum_{i=1}^{c} A_i(\mathbf{x})p_i(y) \tag{8.21}$$

8.9 The Development of the Output Information Granules with the Use of the Principle of Justifiable Granularity

For any input information granule, we form the corresponding information granule B based on some experimental evidence (data) coming in the form of a collection of a one-dimensional (scalar) numeric data (output data) $D = \{y_1, y_2, ..., y_N\}$. Recall that this data set is implied by (conditioned) the input information granule. In this sense, for A_i, the corresponding output information granule is built around such y_{k0} such that $k_0 = \arg\max_{k=1,2,...,N} u_{ik}$ and in the determination of the optimal bounds of the information granule involves the membership grades u_{ik}. For instance in the calculations of the upper bound, we maximize the expression $V(b) = (\sum_{k:m<y_k \leq b} u_k)^* \exp(-\alpha|m-b|)$.

The performance of the granular mapping built upon the family of the granules (A_i, B_i) is quantified in terms of the coverage criterion

$$Q = \frac{\sum_{k=1}^{N} \text{incl}(y_k, Y_k)}{N} \tag{8.22}$$

where $Y_k = \sum_{i=1}^{c} A_i(\mathbf{x}) \otimes B_i$, that is, $[y^-, y^+] = [\sum_{i=1}^{c} A_i(x)b_i^-, \sum_{i=1}^{c} A_i(x)b_i^+]$ along with the associated global assessment $AUC = \int_0^1 Q(\alpha)d\alpha$.

The specificity of the resulting information granule serves as another useful indicator. The higher the specificity of the information granule Y_k is, the more desirable the mapping is. The average length of the information granule Y_k (in case it is realized as an interval)

$$V = \frac{\sum_{k=1}^{N} |y_k^+ - y_k^-|}{N} \tag{8.23}$$

is used to express the specificity (spec) by taking the following expression

$$\text{spec} = 1 - \frac{V}{V(\alpha = 0)} \tag{8.24}$$

Note that the specificity measure, spec, attains the values positioned in-between $[0,1]$ with the higher values pointing at the more preferred character of the mapping from the perspective of its specificity.

As these two measures quantify at a certain conflict (high coverage entails an undesired effect of excessively broad and hence noninformative information granules Y), it becomes beneficial to construct a plot of relationship coverage (Q)–specificity (spec), analyze a shape of the curve, Figure 8.18,

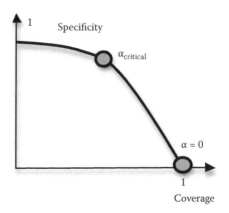

FIGURE 8.18
Plot of coverage versus specificity with a knee point identifying a sound pair (coverage, specificity) associated with a certain value of α. The location of the values of α is displayed in the figure showing a direction of increase in the values of this parameter.

obtained in this way, and eventually identify points at which some sound compromise could be attained. For instance, a location of the critical value of α shown in this figure stresses that going toward higher values of α does substantially reduce the coverage criterion while minimally improving the specificity of the obtained information granules.

8.10 Interpretation of Granular Mappings

Assuming a certain acceptable value of α (being a result of a compromise reached between the coverage and specificity), a granular mapping is interpreted as a collection of rules. To arrive at the rules, the information granules (fuzzy sets) formed in the input space are interpreted by looking at individual variables and associating the prototypes with their semantics articulated for the individual input variables. Consider the prototypes $\mathbf{v}_1$, $\mathbf{v}_2,..,\mathbf{v}_c$. We project them on the corresponding inputs. For the j-th input variables, we obtain the projected prototypes $v_{1j}, v_{2j},...,v_{cj}$. They can be ordered linearly with the labels indexed by successive integers. For each of the integer values we associate a linguistic label along with its numbering, that is, *negative large* (1), *negative medium* (2), *negative small* (3), *around zero* (4), *positive small* (5), *positive medium* (6), *positive large* (7), *very large* (8), and so forth. Proceeding with this labeling process for each variable, the prototype can be represented as an n-tuple of integers $(i_{11}, i_{12}, ...i_{1n})$ for prototype 1, $(i_{21}, i_{22}, ...i_{2n})$ for prototype 2, ... and $(i_{c1}, i_{c2}, ...i_{cn})$ for prototype c. With each of them there is an information granule B_1, B_2, ..., B_c so altogether the granular mapping (or the representation of the input–output data) comes as a collection of rules

$$(i_{11}, i_{12}, \ldots i_{1n}) \rightarrow B_1, (i_{21}, i_{22}, \ldots i_{2n}) \rightarrow B_2 \ldots (i_{c1}, i_{c2}, \ldots i_{cn}) \rightarrow B_c$$

8.11 Illustrative Examples

The experiments reported in this section are focused on the granular mapping realized with the aid of the interval form of information granules formed in the output space. The information granules built on the input space are those obtained by means of fuzzy clustering, FCM, to be precise. The clustering is run for 100 iterations (this number was found to be sufficient when it comes to the convergence of the method; it was found that in all experiments the convergence was reported for, the number of iterations is far below the limit imposed here). This realization of the mapping offers clarity of the presentation and stresses the main points of the construct. The realization of the alternative ways of information granulation could be completed in a similar fashion by admitting the different character of the output information granules.

The detailed parametric studies in which we evaluate the performance of the mapping and its features concern three data sets coming from the Machine Learning Repository (http://archive.ics.uci.edu/ml/) namely, auto, concrete, and housing data. It particular, the evaluation of the mapping is concerned with the number of information granules and its impact on the performance of the mapping expressed in terms of its coverage and specificity (length of the output of the granular mapping). This will be helpful in assessing sound tradeoffs between specificity and coverage vis-à-vis the details of the mapping expressed in terms of the number of input information granules.

Auto data. The granular mapping is realized for the number of input information granules (clusters) varying between two and six. We start with the plots of the coverage and the average length of the resulting information granules treated as functions of α, Figure 8.19. These two plots include these relationships for the varying number of information granules (clusters) in the input space. The points lying on the curves are indexed by the successive values of α.

The plots of the coverage–specificity relationships are displayed in Figure 8.20 for selected values of "c." Apparently there is a visible *knee* point with which a certain value of α can be identified. Here, its optimal value is positioned around 0.4.

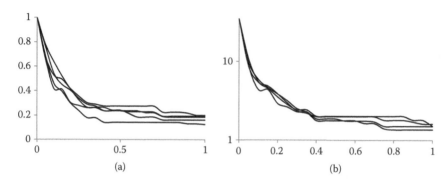

FIGURE 8.19
Coverage (a) and average length of information granule (interval) (b) both being regarded as functions of α.

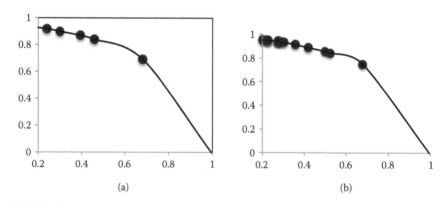

FIGURE 8.20
Coverage–specificity relationships indexed by the values of α and $c = 2$ (a) and $c = 6$ (b). Note a different distribution of the values of α.

The AUC values describing the overall quality of the mapping in terms of the coverage criterion are listed in Table 8.1. The optimal result is obtained for six information granules. While a general tendency is visible (increase in the AUC values occurs with the increase of the number of information granules), there is a departure from this trend at $c = 5$.

TABLE 8.1

AUC Values versus the Number of Information Granules (c)

c	2	3	4	5	6
AUC	0.262	0.335	0.340	0.319	**0.364**

Note: The best result shown in **boldface**.

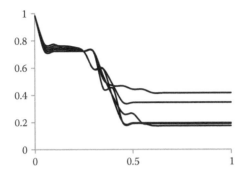

FIGURE 8.21
Plots of coverage versus α for several values of "c."

The dependencies forming the granular mapping for c = 6 and α = 0.1 come in the form

input	output
(2, 2 2 2 2 2 5)	B_1 = [23.84 26.00 28.06]
(5 5 5 5 5 4 2)	B_2 = [15.98 18.00 21.05]
(3 3 3 3 3 6 3)	B_3 = [24.95 28.00 32.15]
(6 6 6 6 1 1 1)	B_4 = [9.98 13.00 16.02]
(1 1 1 1 6 5 6)	B_5 = [30.99 32.40 34.10]
(4 4 4 4 4 3 4)	B_6 = [18.98 23.00 26.07]

(the integer indexes shown in the input space are related to the arrangements of the projections of prototypes on the corresponding input variables as elaborated in Section 8.10).

Concrete data. As before we report the results in terms of the coverage treated as a function of α (see Figure 8.21). Again increasing values of α give rise to the lower values of coverage and several significant drops in the values of coverage associated with the varying values of α are observed.

The AUC values, Table 8.2, indicate the best results reported for c = 5. Again the relationship AUC = AUC(c) does not exhibit a strict monotonic behavior;

TABLE 8.2

AUC Values versus the Number of Information Granules (c)

c	2	3	4	5	6
AUC	0.528	0.428	0.439	**0.577**	0.418

Note: The best result shown in **boldface**.

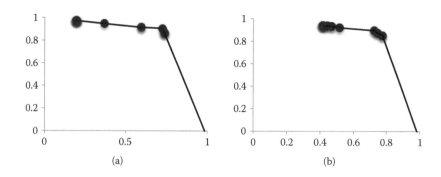

FIGURE 8.22
Coverage–specificity relationships indexed by the values of α and c = 3 (a) and c = 5 (b). Note a different distribution of the values of α as well as the ranges assumed by the coverage measure.

there is the best number of information granules leading to the highest coverage results.

The coverage–specificity plot visualizes the dependency between these two requirements, Figure 8.22. On their basis, one can identify a knee point and determine the most suitable value of α.

Housing data. For this data set, we again report the results in terms of the coverage values, the AUC characterization of the coverage of the granular mapping, and the coverage–specificity dependencies, Figure 8.23. While the trends are similar as reported for the two previous data sets, it is visible that the AUC is not monotonic with regard to the number of information granules (it rather seems to be the most suitable number of information granules that are able to deliver the best representation of the structure in the data, which becomes reflected in the highest value of the AUC), Table 8.3. Again the knee points on the coverage–specificity graphs are also clearly visible, Figure 8.24.

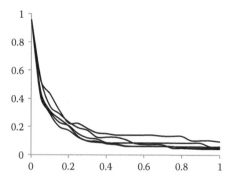

FIGURE 8.23
Plots of coverage versus α for several values of "c."

TABLE 8.3

AUC Values versus the Number of Information Granules (c)

c	2	3	5	7	9
AUC	0.177	**0.234**	0.189	0.165	0.157

Note: The best result shown in **boldface.**

8.12 Conclusions

The methodological underpinnings of granular fuzzy models stem from the need of an effective conceptual and algorithmic representation and quantification of diversity of locally available sources of knowledge (detailed models). The diversity of views at the problem/system/phenomenon is quantified via the granularity of results produced by the global model constructed at the higher level of hierarchy. The granular nature of the results formed there is inherent to the diversity of the sources of knowledge.

The quantification of granularity itself (namely, the multifaceted nature of available models) is a direct result of multiobjective optimization (Ishibuchi and Nojima, 2007)—it is shown that the criteria of coverage of data and specificity of information are conflicting in nature. The choice of a suitable tradeoff in the satisfaction of these two requirements is left to the user. Nevertheless, the AUC is helpful here as it can quantify an overall performance of the global model and rank several global models through the use of the AUC values.

Granular fuzzy models subsume the concept of type-2 fuzzy models in the sense that they offer compelling, algorithmically well-supported evidence behind the emergence of fuzzy models of higher type. They are more general than type-2 fuzzy models as here we are not confined to any particular

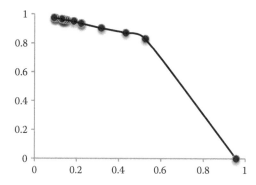

FIGURE 8.24

Coverage–specificity relationships indexed by the values of α and $c = 3$.

architecture of fuzzy models and the way in which type-2 fuzzy sets are incorporated into specific components of the models.

It is worth noting that the notion of granular modeling stretches beyond the framework of fuzzy models. In essence, as we deal with models articulating locally available knowledge, the quantification of the diversity of such sources becomes encapsulated through information granules produced by granular models. In this way, we can talk about *granular* neural networks (in the case of local models being formed as neural networks), *granular* regression (when dealing with individual regression models), or *granular* fuzzy cognitive maps (when the local models are fuzzy cognitive maps), and so forth. One can look into further generalizations in the form of granular fuzzy models of higher type, that is, type-2 granular fuzzy models or *granular*^{2} fuzzy models. Those are a result of dealing with several hierarchies of sources of knowledge, namely, fuzzy models formed in a two-level hierarchical architecture of the knowledge sources.

References

Alcala, R., P. Ducange, F. Herrera, B. Lazzerini, and F. Marcelloni. 2009. A multiobjective evolutionary approach to concurrently learn rule and databases of linguistic fuzzy-rule-based systems. *IEEE Trans. on Fuzzy Systems*, 17, 1106–1122.

Hastie, T., R. Tibshirani, and J. Friedman. 2009. *The Elements of Statistical Learning: Data Mining, Inference, and Prediction*, 2nd ed. Berlin: Springer.

Ishibuchi, H. and Y. Nojima. 2007. Analysis of interpretability-accuracy tradeoff of fuzzy systems by multiobjective fuzzy genetic-based machine learning. *Int. J. of Approximate Reasoning*, 44, 4–31.

Klir, G. and B. Yuan. 1995. *Fuzzy Sets and Fuzzy Logic: Theory and Applications*. Upper Saddle River: Prentice-Hall.

Pedrycz, W., V. Loia, and S. Senatore. 2010. Fuzzy clustering with viewpoints. *IEEE Transactions on Fuzzy Systems*, 18, 274–284.

Pedrycz, W. and P. Rai. 2008. Collaborative clustering with the use of Fuzzy C-Means and its quantification. *Fuzzy Sets and Systems*, 159, 2399–2427.

Pham, D.T. and M. Castellani. 2006. Evolutionary learning of fuzzy models. *Engineering Applications of Artificial Intelligence*, 19, 583–592.

Roubos, H. and M. Setnes. 2001. Compact and transparent fuzzy models and classifiers through iterative complexity reduction. *IEEE Trans. on Fuzzy Systems*, 9, 516–524.

Zadeh, L.A. 1975. The concept of linguistic variables and its application to approximate reasoning I, II, III. *Information Sciences*, 8, 199–249, 301–357, 43–80.

Zadeh, L.A. 1997. Towards a theory of fuzzy information granulation and its centrality in human reasoning and fuzzy logic. *Fuzzy Sets and Systems*, 90, 111–117.

9

Granular Time Series

In this chapter, we introduce a concept of granular time series—models of time series in which information granules play a central role. Information granularity is instrumental in arriving at human-centric models of high level of interpretability and transparency. We consider description, interpretation, and classification of time series by showing how a layered architecture of granular time series is developed. In the sequel, a granular classifier of time series with a granular feature space is discussed. The interpretability of time series is of significant interest in data mining and the granular time series makes a certain contribution to this area.

9.1 Introductory Notes

In spite of the remarkable diversity of models of time series, there is still a burning need to develop constructs, whose accuracy and interpretability are carefully reconciled and balanced thus leading to highly interpretable (human-centric) models. While a great deal of research has been devoted to the design of nonlinear models, that is, neural networks, of time series (with an evident expectation of achieving high accuracy of prediction), an issue of interpretability (transparency) of the models of time series becomes an evident and ongoing challenge. The user-friendliness of models of time series comes hand in hand with the ability of humans to perceive and process abstract entities rather than plain numeric entities. In the perception of time series, information granules regarded as such entities play a pivotal role thus giving rise to granular models of time series or *granular* time series, in brief. In the proposed model, we elaborate on the fundamental hierarchically organized layers of processing supporting the development of granular time series, namely, (a) granular descriptors used in the visualization of time series, (b) construction of linguistic descriptors used afterward in the generation of (c) a linguistic description of time series. The layer of linguistic prediction models of time series exploiting the linguistic descriptors is the one formed at the highest level.

9.2 Information Granules and Time Series

The pursuits of perception, analysis, and interpretation of time series, as realized by humans, are completed at a certain, usually problem-implied, level of detail. Instead of single numeric entries—successive *numeric* readings of time series, formed are conceptual entities over time and feature space—information granules using which one discovers meaningful and interpretable relationships forming an overall description of time series. The granularity of information is an important facet imperative to any offering of well-supported mechanisms of comprehension of the underlying temporal phenomenon. In all these pursuits, information granules manifest along the two main dimensions (as noted above). The first one is concerned with time granularity. Time series is split into a collection of time windows—temporal granules. One looks at time series in temporal windows of months, seasons, and years. Time series is also perceived and quantified in terms of information granules formed over the space of amplitude of the successive samples of the sequence; one arrives at sound and easily interpretable descriptors such as *low, medium, high* amplitude and the like. One can form information granules over the space of changes of the time series. Combined with the temporal facets, the composite information granules arise as triples of entities, say *long* duration, positive *high* amplitude, *approximately* zero changes, and so forth. Once information granules are involved, long sequences of numbers are presented in far shorter, compact sequences of information granules—conceptual entities as noted above, which are easier to understand and process. Let us note that while temporal granules are quite common (and they are of the same length), forming and processing of composite information granules still calls for detailed investigations.

9.3 A Granular Framework of Interpretation of Time Series: A Layered Approach to the Interpretation of Time Series

As noted so far, the notion of information granularity plays a pivotal role in all interpretation and analysis pursuits of time series. Our investigations of the description of time series being cast in the setting of Granular Computing are presented in a top-down manner. We start with an overall view of the conceptual framework stressing its key functionalities and a layered architecture and then move on to a detailed discussion by elaborating on the supported algorithmic aspects.

As commonly encountered, a starting point is numeric data—time series $\{x_1, x_2, ..., x_N\}$. For the purpose of further analysis, we also consider the dynamics of the time series by considering the sequences of differences (changes), namely, $\{\Delta x_2, \ldots, \Delta x_N\}$ where $\Delta x_i = x_i - x_{i-1}$. The space in which time series are discussed comes as a Cartesian product of the space of amplitude $\mathbf{X}$ and change of amplitude, namely, $\mathbf{X} \times \Delta \mathbf{X}$.

A bird's $1 = 1$ eye view of the overall processing stressing the associated functionality is displayed in Figure 9.1.

Let us elaborate in more detail on the successive layers at which consecutive phases of processing are positioned:

Formation of information granules. This formation is over the Cartesian product of amplitude and change of amplitude $\mathbf{X} \times \Delta \mathbf{X}$. Here for each time slice T_i (note that the time is subject to granulation as well), we form an interval information granule (Cartesian product) $X_i \times \Delta X_i$. These information granules are constructed following the principle of justifiable granularity (Section 5.1 in Chapter 5). At this phase, the temporal granulation is optimized (one should stress that as not all time slices are of the same length, some optimization mechanism has to be considered here). An overall volume of all information granules is used here as an optimization criterion.

Visualization of information granules. The results produced at the previous processing phase are visualized. In essence, for each time slice (segment) T_i, one can produce a contour plot of the Cartesian product of the information granule in $\mathbf{X} \times \Delta \mathbf{X}$.

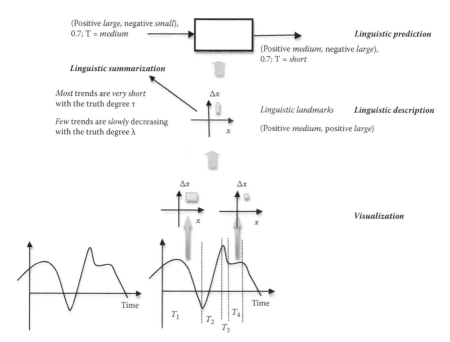

FIGURE 9.1
Development of granular time series: visualization and linguistic interpretation. Emphasized is a layered hierarchical approach presented in this study.

Linguistic description of granular time series. While the visualization of the granular time series could be quite appealing, it is also worth arriving at the linguistic description of the time series, which is articulated in terms of some linguistic (granular) landmarks and levels of the best matching of these landmarks with the information granules $X_i \times \Delta X_i$. Having a collection of the landmarks A_I, $I = 1, 2, \ldots$, c where typically c $\ll$ p, the linguistic description comes as a string of the landmarks (for which the best matching has been accomplished) along with the corresponding matching levels λ_1, $\lambda_2, \ldots, \lambda_p$. For instance, the linguistic description can read as the following sequence of descriptors,

$$\{(\text{positive } small \times \text{negative } medium) \ (0.7) \ T_1\}$$

$$\{(\text{negative } large \times around \text{ zero}) \ (0.9) \ T_2\} \ \ldots$$

where each granular landmark comes with its own semantics, (e.g., *positive small*, *medium*, and so on) in virtue of the space in which the time series is described (amplitude and its change). The triples linguistically describe the amplitude and its change (expressed in terms of linguistic terms), the associated matching level, and the duration of the time slice (T_j).

To arrive at the linguistic description of this nature, two associated tasks are to be handled, namely, (a) a construction of meaningful (semantically sound) granular landmarks, and (b) invoking a matching mechanism, which returns a degree of match achieved. The first task calls for some mechanism of clustering of information granules while the second one is about utilizing one of the well-known matching measures encountered in fuzzy sets, that is, possibility or necessity measures.

Linguistic prediction models of time series. The descriptions of the time series are useful vehicles to represent (describe) time series in a meaningful and easy to capture way. Per se, they are not models such as standard constructs reported in time series analysis. They do, however, deliver all components, which could be put together to form granular predictive models. Denoting by $A_1, A_2, \ldots, A_c$ the linguistic landmarks developed at the previous phase of the overall scheme, a crux of the predictive model is to determine relationships between the activation levels of the linguistic landmarks for the current time granule T_k and those levels in the next time granule T_{k+1}. The underlying character of the predictive mapping can be schematically expressed in the following way,

$$A_1(X_k), A_2(X_k), \ldots, A_c(X_k) \rightarrow A_1(X_{k+1}), A_2(X_{k+1}), \ldots, A_c(X_{k+1}) \qquad (9.1)$$

where $A_i(X_k)$ stands for a level of activation (matching) between A_i and the current information granule X_k. The operational form of the predictive model can be produced in the form of a granular relational equation

$$\mathbf{A}(X_{k+1}) = \mathbf{A}(X_k) \circ R \qquad (9.2)$$

with "∘" being a certain composition operator (i.e., max-min or max-t composition) completed over information granules. $A_i(X_{k+1})$ is the activation level of the linguistic landmarks A_i by the predicted information granule X_{k+1}. Overall, the vector $A(X_k)$ has "c" coordinates, namely, $[A_1(X_k)\ A_2(X_k)$ ... $A_c(X_k)]$. The granular relation R of size $c*c$ captures the relationships between the corresponding entries of $A(X_k)$ and $A(X_{k+1})$. Note that X_{k+1} is not specified explicitly, but rather through the degrees of activation of the linguistic landmarks. In other words, the predictive granular model returns a collection of quantified statements (the corresponding entries of $A(X_{k+1})$)

—predicted information granule is A_1 with degree of activation $\lambda_1\ (= A_1(X_{k+1}))$
—predicted information granule is A_2 with degree of activation $\lambda_2\ (= A_2(X_{k+1}))$

$$\cdots$$

—predicted information granule is A_c with degree of activation $\lambda_c\ (= A_c(X_{k+1}))$

$$(9.3)$$

which offer a certain intuitive view of the prediction outcome. Obviously, one can easily choose the dominant statement for which the highest level of matching has been reported, namely,

$$\text{—predicted information granule is}$$
$$A_i \text{ with degree of activation } \lambda_i\ (= A_i(X_{k+1})) \tag{9.4}$$

where

$$\lambda_i = \arg\max_{j=1,2,\ldots,c} A_j(X_{k+1}) \tag{9.5}$$

Note that there might be also another relational dependency for the predicted next time granule T_{k+1}

$$A_1(X_k), A_2(X_k), \ldots, A_c(X_k), T_k \to T_{k+1} \tag{9.6}$$

which can be regarded as a certain granular relational equation

$$T_{k+1} = A(X_k) \circ T_k \circ G \tag{9.7}$$

We can regard these predictive models Equation (9.7) as granular models of order 2 as they are described over the space of information granules rather than numeric entries.

Linguistic summarization. This is one of the possible realizations of user-centric tasks, which is based on the linguistic granules. As proposed in Kacprzyk and Zadrozny (2005) and Kacprzyk, Wilbik, and Zadrozny (2006), the objective is to construct summaries of the time series based on a collection of information granules. Here, typical examples include description in the form quantified with some truth degrees (λ, μ, τ),

> *Among* all increasing segments, most of them are *short*
>
> *Most* trends are *very short* with the truth degree τ
>
> *Few* trends are *slowly* decreasing with the truth degree λ
>
> *Over* half trends with a *high* variability are constant with the truth degree μ

where *most, usually, few, among all,* and so forth are so-called linguistic quantifiers (Kacprzyk and Yager, 2001). Note that the summaries of this nature offer a high-level view of the time series not including temporal dependencies but rather focused on the high-level view of the temporal data.

9.3.1 Formation of Interval Information Granules

For the data contained in the time interval T_i, we consider samples of time series $\{x_1, x_2, ..., x_N\}$ and their differences $\{\Delta x_1, \Delta x_2, ..., \Delta x_{N-1}\}$ out of which, following the principle outlined in Section 3.1 in Chapter 3, we construct interval information granules X_i and ΔX_i and their Cartesian product. As these Cartesian products are indexed by the values of α, we form a fuzzy set—a family of α-cuts $(X_i \times \Delta X_i)_\alpha$.

9.3.2 Optimization of Temporal Intervals

As noted, the time intervals T_i over which information granules are not of the same length (which is quite intuitively appealing as the variability of the time series changes over time and the information granules formed over these time intervals). To optimize the lengths of the intervals, we formulate an optimization problem. First, we introduce a volume of information granule by computing the length of it (T_i) by the area of the fuzzy set (α-cuts) $(X_i \times \Delta X_i)_\alpha$ and sum them over all time intervals obtaining V. This total volume is minimized over $T_1, T_2, ..., T_c$, that is, min $_{T1, T2,..., Tc}$ V.

In the optimization of the total volume, we can resort to some methods of global optimization such as, for example, particle swarm optimization, genetic algorithms (GAs), artificial bee colonies (ABC), or the like. Also a stepwise realization of the temporal intervals could be sought in which one determines only two time intervals once at a time by finding the minimum of V over T_1 (which results in a single variable optimization problem) and then proceeds with successive refinements by splitting one of the time intervals for which the value of V is higher. The process is repeated in the same way when forming further refinements. For comparative reasons, it is instructive to compare the total volume of information granules for a uniform distribution of the cutoff points (temporal windows of equal size).

9.3.3 Clustering Information Granules—A Formation of Linguistic Landmarks

Once we have a collection of information granules, they are clustered. The algorithm of fuzzy c-means is one viable alternative. For the predetermined number of clusters, that is, "c," we form a collection of (granular) prototypes $V_1 \times \Delta V_1, V_2 \times \Delta V_2, ..., V_c \times \Delta V_c$. Schematically, we can illustrate the process in Figure 9.2.

As the objects to be clustered are information granules, not vectors of real numbers, the space in which clustering is realized has to be carefully organized. A viable option is to look at the description of the information granules as a collection of several α-cuts. By doing this, for the purpose of clustering, we form an augmented feature space. If "p" α-cuts were selected, the dimensionality of the space in which clustering takes place is increased "p" times so one needs to be careful and consider only a limited number of these α-cuts.

Alternatively, one can describe information granules in some parametric form and proceed with clustering in this new feature space. For the representation issues, refer to some studies reported in Hathaway, Bezdek, and Pedrycz (1996) and Pedrycz et al. (1998).

9.3.4 Matching Information Granules and a Realization of Linguistic Description of Time Series

For the linguistic description of granular time series, it becomes essential to match the fuzzy relations $X_i \times \Delta X_i$ (their α-cuts) with the prototypes obtained during fuzzy clustering. The essence is to complete matching of $X_i \times \Delta X_i$ and the prototype $V_I \times \Delta V_I$ by computing the possibility measure

Amplitude: NL, NM, NS, Z, PS, PM, PB

Change of amplitude: NL, NM, NS, Z, PS, PM, PB

Time interval: S, M, L, VL

Formation of clusters (landmarks) over information granules and time intervals

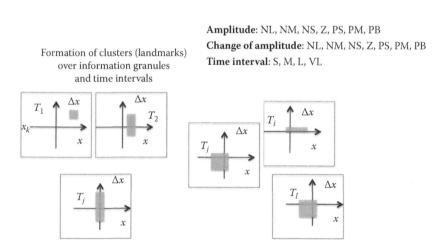

FIGURE 9.2
From granular data to linguistic landmarks.

Poss($X_i \times \Delta X_i$ and $V_I \times \Delta V_I$). These calculations of possibility are straightforward as we find a maximal value of α for which the corresponding α-cuts of $X_i \times \Delta X_i$ and $V_I \times \Delta V_I$ overlap. We repeat the calculations for all $I = 1, 2, ..., c$ and choose I_0 with the highest possibility measure. Denote it by λ_i. The process is carried out for successive information granules $X_i \times \Delta X_i$ giving rise to a series of linguistic descriptors along with their levels of matching,

$$(V_{I1} \times \Delta V_{I1}, \lambda_1)\ (V_{I2} \times \Delta V_{I2}, \lambda_2)... (V_{Ip} \times \Delta V_{Ip}, \lambda_p)$$

9.4 A Classification Framework of Granular Time Series

The schemes of granular classification are comprised of several functional modules and while there are some similarities with the commonly encountered schemes of pattern classification, we encounter here several significant differences. A crux of the overall scheme proposed in this study and displayed in Figure 9.3 can be captured schematically as follows:

data → feature space → granular feature space → interpretation → classifier

Let us briefly elaborate on the underlying functionality of the modules comprising the overall scheme. The highlights of the main functional phases help understand the very nature of the processing; to clarify a pivotal role information granularity plays as a part of the scheme both in terms of the facilitation of processing itself as well as formation of a vehicle to facilitate interaction with the user.

Let us briefly elaborate on the essence of the successive phases of the overall processing scheme. This will help us stress a systematic and coherent development process being proposed in this study as well as highlight its novel facets along with the role and motivation behind the ensuing information technologies exploited in the process.

Representation of time series—a formation of the feature space. There is a remarkable diversity of representation schemes for time series including those benefiting from the techniques originating from statistics, computational intelligence, and data analysis. In general, one can allude to time series representations completed in the temporal domain (i.e., autoregressive [AR],

FIGURE 9.3
Granular time series description and classification: an overview of processing.

autoregressive moving average [ARMA] models, neural networks, wavelets) (Aznarte and Benitez, 2010) as well as those carried out in the frequency domain (e.g., spectral descriptors of time series). A lot of recent studies reported in the literature serve as a testimony to these developments. In spite of the inherent diversity, these representations share a striking commonality: they are predominantly of numeric character. As a result of this representation, one returns a vector of numeric descriptors characterizing the time series and used in consecutive phases of analysis, description, and classification. The vectors of numeric descriptors form a feature space within which the time series is positioned. As usual in pattern recognition, our anticipation is that the feature space formed with the use of carefully selected features delivers significant discriminatory properties, which help delineate patterns (time series) belonging to different classes.

Granular representation of time series. In this phase of the overall process, we design a collection of information granules in the feature space already formed as a result of the characterization of time series done at the first phase of the scheme. The intention is to gain a holistic bird's-eye view of the structure of a collection of time series (and realized with the use of a given description). Information granules are designed in various ways, however, clustering or fuzzy clustering along with their numerous variants accommodating some mechanisms of supervision (as limited labeling of time series could be available) are commonly considered here. The granular representation of time series comes with useful interpretation abilities: the clusters (information granules) can be assessed with regard to their class homogeneity and this could help choose a suitable level of granularity and/ or make a decision on an alternative realization of the feature space. The use of information granules is also useful in the reduction of the granular feature space for which granular classifiers are formed. In most situations, we arrive at the characterization of clusters in terms of their prototypes and partition matrices.

Granular classifiers. The classifiers positioned as the last functional module of the scheme are used to realize mapping of time series on class labels. Considering the more abstract nature of information granules, the discriminatory mapping of the classifier is conveniently realized as a logic-inclined mapping between the activation levels of the information granules (membership functions in the case of fuzzy sets). The constructs falling under this rubric offer the potential for transparency of logic nature of the mapping associated with its nonlinear character supporting classification abilities.

It is worth stressing the diversity of possible feature spaces as well as induced granular feature spaces (implied by the diversity of information granules as well as the assumed level of information granularity itself). Several families of detailed topologies of the granular classification schemes can be envisioned as outlined in Figure 9.4.

The highly visible aspect of the alternatives comes with a diversity of feature spaces and the granular feature spaces available in the proposed architecture. Referring to Figure 9.4, some alternatives are worth highlighting:

(a) A single feature space follows several granular feature spaces, which entails that the same representation of time series is looked at and used for classification purposes. In particular, by varying the number of information granules, several views of the feature space exhibiting different levels of detail are exploited for classification purposes. The underlying concept is visualized in Figure 9.4a.

(b) As shown in Figure 9.4b, several feature spaces are considered and with each of them is associated a certain granular feature space.

(c) The structure outlined in Figure 9.4c builds upon the two schemes discussed above: a number of feature spaces are formed and each of them gives rise to several granular feature spaces.

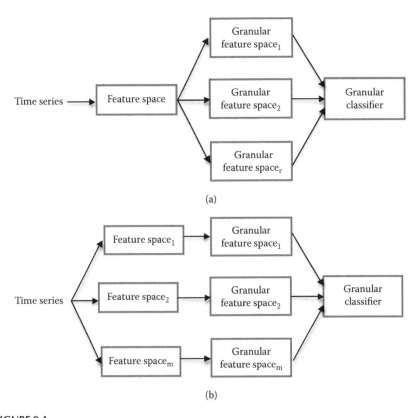

FIGURE 9.4
Selected alternatives in the representation of time series. (a) The same feature space associated with various granular feature spaces. (b) Different feature spaces associated with granular feature spaces. (*Continued*)

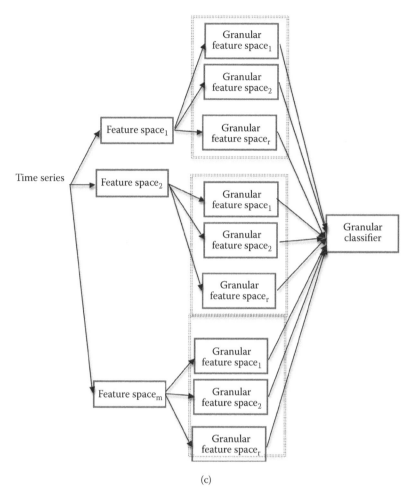

(c)

FIGURE 9.4 *(Continued)*
Selected alternatives in the representation of time series. (c) Different feature spaces where for each of them is associated a collection of granular feature spaces.

One has to emphasize that the granular description of temporal data and information granules (clusters) formed there play an essential role in the overall processing pursuits. Information granules that can be regarded as more abstract and interpretable entities forming a new granular feature space and classification pursuits are carried out in this new space. Not too much research has been completed so far and the entire area is open for vigorous investigations. Here we may encounter a great deal of originality as a number of fundamental questions about the process and an assessment of the resulting information granules have not been posed and need to be carefully addressed. In the formation of information granules and the overall granular representation space, we have to look at both aspects of discriminatory capabilities of information granules. Let us note that owing to the nonlinearity furnished by information granules (i.e., in terms of nonlinear membership functions of fuzzy sets or characteristic functions of sets), the performance of classification schemes (classifiers) could be significantly enhanced if such nonlinear properties were prudently exploited. Likewise the dimensionality of the space in which the granular classifier is realized is affected by the space of information granules.

A few lines of comparison between the proposed topology of the processing scheme and pattern classifiers commonly investigated in the realm of pattern classification could be of help here in order to cast the problem discussed here in a certain setting. The striking difference is that the classifier is constructed not in the feature space but the space formed by information granules. The logic-oriented character of the mapping realized by the classifier (operating on activation levels of information granules) is also a visible feature of the overall construct.

In what follows, we now proceed with a detailed description of the successive phases of the general classification scheme.

9.4.1 Building a Feature Space for Time Series Representation and Classification

Current research in time series description, and time series characterization and prediction comes with a visible diversity and richness of the conceptual and algorithmic pursuits. For the purposes of reduction of data and facilitating all mining algorithms, in most development schemes discussed is a segmentation step. Time series segmentation can be treated either as a preprocessing stage for a variety of ensuing data mining tasks or as an important stand-alone analysis process. An automatic partitioning of a time series into an *optimal* (or better to say, *feasible*) number of homogeneous temporal segments becomes an important problem. Quite often considered is a fixed-length segmentation process. Common segmentation methods include the use of the perceptually important points (PIP) in time series as the time points, minimum message length (MML), and minimum description length (MOL) or (MDL) segmentation (Oliver, Bexter, and Wallace, 1998). A two-stage approach, in which one first uses piecewise generalized likelihood ratio (GLR) to carry out rough segmentation and then refines the results, has been proposed by Wang and Willett (2004). Keogh et al. (2001) adopted a piecewise linear representation (PLR) method to segment time series. They focused on the problem of an online segmentation of time series where a sliding window and bottom-up (SWAB) approach was proposed. Fuzzy clustering algorithms have showed a significant potential to address this category of problems.

With this regard, Abonyi et al. (2005) developed a modified Gath–Geva (GG) algorithm to divide time-varying multivariate data into segments by using fuzzy sets to represent temporal segments. By using dynamic programming, one determined a total number of intervals within the data, the location of these intervals, and the order of the model within each segment. In the segmentation problem, one considered a tool for exploratory data analysis and data mining called the *scale-sensitive gated experts* (SSGE), which can partition a complex nonlinear regression surface into a set of simpler surfaces called *features*. A recent survey of mining time series is covered by Fu (2011).

9.4.2 Formation of a Granular Feature Space

As a result of clustering, we obtain "c" clusters which are fully described by the corresponding prototypes $\mathbf{v}_1, \mathbf{v}_2, \ldots, \mathbf{v}_c$ or equivalently fuzzy sets $A_1, A_2, \ldots, A_c$ whose membership functions are computed as follows:

$$A_i(\mathbf{z}) = \frac{1}{\sum_{j=1}^{c} \left(\frac{||\mathbf{z}-\mathbf{v}_i||}{||\mathbf{z}-\mathbf{v}_j||} \right)^{2/(m-1)}} \tag{9.8}$$

Any vector representing time series in a certain representation space (feature space), that is, $\mathbf{z}$, results in a collection of membership degrees, $\mathbf{x} \in [0,1]^c$ where the membership degrees are computed by looking at the closeness of $\mathbf{z}$ with regard to the prototypes.

The quality of the granulation–degranulation process being realized with the aid of the clusters is evaluated.

Let us stress that the granulation of the initial (numeric) representation space exhibits two interesting aspects. First, granulation could reduce the dimensionality of the problem. The space formed here is of dimensionality "c" where typically c < n. Furthermore, the process results in a nonlinear transformation of the original feature space (observe that Equation 9.8 provides a nonlinear transformation of the numeric representation of signals), which could enhance discriminatory capabilities of the ensuing classifier.

There are two main approaches to the formation of the information granules in the granular feature space:

(a) As fuzzy clustering is a method of unsupervised learning, one can look at the data (representations of the series) as not carrying any class labels and form the clusters. Obviously, the number of clusters (c) has to be equal or greater than the number of classes.

(b) Clustering can be realized for time series belonging to the individual classes. Here we consider time series belonging to a given class.

9.5 Granular Classifiers

In this section, we elaborate on a design process of the granular classifier starting with a discussion on its architecture (where several alternatives are investigated and provided with their interpretation) and presenting various ways in which the parameters (the entries of the fuzzy relation) of the classifier are determined.

9.5.1 Underlying Architecture of the Classifier

Owing to the nature of the granular feature space, we consider here a relational category of classifiers being realized in the form of relational mappings between information granules and class assignment. They offer an interesting aspect revealing and capturing logic relationships that are dominant to the corresponding classes. Assuming that information granules (fuzzy sets) forming two granular feature spaces are denoted as A_i, and B_j, for the corresponding spaces, the relational dependency between the class membership and the associated activation (compatibility) levels can be described in the form

$$\omega = (A_i \times B_j) \text{ op } R \tag{9.9}$$

where ω is a p-dimensional vector of class membership (note that in virtue of using fuzzy sets, we may encounter degrees of membership to individual classes rather than a Boolean "yes-no" binary quantification of class assignment). Here the symbol "op" denotes a certain relational composition operator, which is used to compose a fuzzy relation of the classifier R with the current granular description of a given time series to be classified while × stands for a Cartesian product of the coordinates of the granular feature space. Let us look at more details in Equation (9.9) by identifying individual variables. Refer also to Figure 9.5.

For a given time series described in a certain feature space and coming as a certain **z**, the resulting vector in the granular feature space is determined on

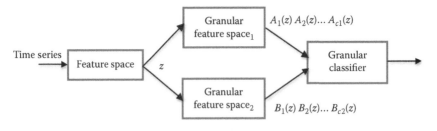

FIGURE 9.5
Realization of the relational classifier in the case of two granular feature spaces formed for a certain feature space.

the basis of the existing prototypes by computing the values of the levels of activation (compatibility) $A_1(z)$, $A_2(z)$, ..., $A_{c1}(z)$ and $B_1(z)$, $B_2(z)$, ..., $B_{c2}(z)$ with c_1 and c_2 being the number of information granules formed in the feature space. The details of the scheme are included in Figure 9.5.

It is convenient to introduce here the following concise vector notation $x = [A_1(z)\ A_2(z)...A_{c1}(z)]$ and $y = [A_1(z)\ A_2(z)...A_{c1}(z)]$. The vector of class membership ω is p-dimensional with "p" coordinates describing class membership. This helps us rewrite Equation (9.9) in the following form:

$$\omega = (x \times y) \text{ op } R \qquad (9.10)$$

We rewrite Equation (9.10) in terms of the individual coordinates of the components. Note that the Cartesian product is modeled by a certain t-norm (or the minimum operation, in particular), $\min(x_i, y_j)$ or $t(x_i, y_j)$. Overall, we have

$$\omega_l = \text{op}_{i,j}[t(x_i, y_j), r_{ijl}] \qquad (9.11)$$

$l = 1, 2, ..., p$ where the aggregation operator realizes a convolution of components of the Cartesian product with the corresponding entries of the relation (matrix) R.

In the above topology of the classifier, we considered two granular feature spaces (see Figure 9.5) to fully illustrate the underlying processing. This description of the granular classifier could be easily scaled down to a single feature space or scaled up to the granular feature space of higher dimensionality (with a number of descriptions).

Several alternatives for the composition operators (op) along with well-articulated logic interpretations will be investigated. In particular, one can investigate two logic-based compositions encountered in fuzzy sets:

(a) s-t or max-min composition of $x \times y$ and R

$$\omega = (x \times y) \circ R \qquad (9.12)$$

which in terms of the individual elements of the vectors reads as follows:

$$\omega_l = \max_{i,j} [\min(\min(x_i, y_j), r_{ijl})] \qquad (9.13)$$

(the min and max operators can be generalized to any t-norm and t-conorm).

(b) t-s or min-max composition of the complement of $x \times y$ and the fuzzy relation R

$$\omega = \overline{x \times y} \bullet R \qquad (9.14)$$

Again we express the above relationship in terms of the membership grades as follows:

$$\omega_l = \min_{i,j} [\max(1 - \min(x_i, y_j), r_{ijl})] \qquad (9.15)$$

As an illustrative example, let us discuss a two-class problem with a single input x in which the relation R has the following entries

[1.0 0.8 0.2 0.1 0.0] for class ω_1 and   [0.2 0.3 0.8 0.9 1.0] for class ω_2.

Consider the input x taking on the form [0.1 1.0 0.3 0.2 0.0]. By carrying out the max-min composition, we obtain class membership vector ω = [0.8 0.3] whereas in the min-max composition (note that here we involve the complement of x, namely, [0.9 0.0 0.7 0.8 1.0]), one has the class membership vector with the entries [0.7 0.3]. Altogether, the obtained class membership vectors quantify that x belongs to the first class with the membership grades positioned in-between 0.7–0.8.

9.5.2 A Construction of the Fuzzy Relation of the Classifier

The fuzzy relation of the classifier can be developed in several ways. We discuss two of them exhibiting a different approach to the design of the classifier.

Gradient-based learning scheme. The gradient-based scheme operates in a supervised mode in presence of some input–output data where the inputs are granular representations of time series x_k, y_k while the output is a binary vector of class membership $\textbf{target}_k$, k = 1, 2, …, N. A performance index quantifies a distance between $\textbf{target}_k$ and ω_k coming from the granular classifier; typically,

$$Q = \sum_{k=1}^{N} (\omega_k - \textbf{target}_k)^T (\omega_k - \textbf{target}_k) \qquad (9.16)$$

The update formula for the gradient-based learning is described concisely as

$$R(\text{iter} + 1) = R(\text{iter}) - a \nabla_R Q \qquad (9.17)$$

where $\nabla_R Q$ is a gradient of Q computed with respect to R and α stands for a positive learning rate and the iteration index (iter) goes from 0, 1, 2,…. An initial fuzzy relation R(0) is the one which accumulated the existing experimental evidence in the form of the union of Cartesian products of input–output data, namely,

$$R(0) = \bigcup_{k=1}^{N} (x_k \times y_k \times \omega_k) \qquad (9.18)$$

The gradient-based schemes are well known in the literature with a variety of applications. A certain disadvantage comes with a form of the minimized performance index, which albeit susceptible to the gradient-based optimization, does not capture the classification error.

Evolutionary optimization of the classifier. Note that in gradient-based learning we minimize Equation (9.17) while the performance of the classifier is expressed in terms of classification rate or any other measure typical for assessing the performance of classification schemes. Those measures cannot be directly involved in the gradient-based optimization, which is a reason why one can resort to a more elaborate optimization scheme.

In the design of the classifier, the gradient-based learning minimizes the performance index Equation (9.16), which, however, is not fully reflective of the quality of the classifier, namely, the classification error (to be minimized) or the classification accuracy (to be maximized). In other words, when minimizing Q, there is no guarantee that the classification error becomes minimized. With this regard, a use of methods of evolutionary optimization is advisable given the flexibility of the fitness function and the nature of the optimization process that is not guided by the gradient information.

To make the fitness function fully reflective of the performance of the classifier (so that the classifier can be effectively optimized), we determine the maximal entry of the vector ω_k for a given Cartesian product $\mathbf{x}_k \times \mathbf{y}_k$ along with its location in the vector of class membership and form a binary vector $\mathbf{b}_k$ with a single entry set to 1 (the others are set to zero) positioned at the j_0-th coordinate where

$$j_0 = \arg{}_{j=1,2,\ldots,p} \max \omega_{kj}. \tag{9.19}$$

The **target**$_k$ is a binary vector with a single entry set to 1. The distance of this vector from $\mathbf{b}_k$ being usually different from zero indicates that the k-th pattern has been misclassified. The sum of these distances (with the summation completed for all patterns) denoted by V is regarded as the fitness function; its minimization is equivalent to the minimization of the classification error. In other words, V is a classification error and its minimization is realized in an explicit manner.

9.6 Conclusions

Granular Computing and granular constructs bring a badly needed facet of models transparency to time series and facilitate effective communication with the user as an active participant of the overall modeling process. Information granules help strike a sound balance between accuracy and transparency (interpretability). Information granules developed both in the feature space in which time series are represented and the time axis are advantageous in forming meaningful descriptors of the temporal data. It is shown that the hierarchical structure plays an important role in the formation of various layers of conceptual components, especially information granules and linguistic descriptors. The level of detail of granular time series is easily adjustable and as a result produces a spectrum of models of variable accuracy.

References

Abonyi, J., B. Feil, S. Nemeth, and P. Arva. 2005. Modified Gath-Geva clustering for fuzzy segmentation of multivariate time-series. *Fuzzy Sets and Systems*, 149, 39–56.

Aznarte, J.L. and J.M. Benitez. 2010. Equivalences between neural-autoregressive time series models and fuzzy systems. *IEEE Transactions on Fuzzy Systems*, 21, 9, 1434–1444.

Fu, T.C. 2011. A review on time series data mining. *Engineering Applications of Artificial Intelligence*, 24, 164–181.

Hathaway, R.J., J.C. Bezdek, and W. Pedrycz. 1996. A parametric model for fusing heterogeneous fuzzy data. *IEEE Trans on Fuzzy Systems*, 4, 270–281.

Kacprzyk, J., A. Wilbik, and S. Zadrozny. 2006. Using a genetic algorithm to derive a linguistic summary of trends in numerical time series. *International Symp. on Evolving Fuzzy Systems*, 137–142.

Kacprzyk, J. and R.R. Yager. 2001. Linguistic summaries of data using fuzzy logic. *Int. Journal of General Systems*, 30, 33–154.

Kacprzyk, J. and S. Zadrozny. 2005. Linguistic database summaries and their proto-forms: Toward natural language-based knowledge discovery tools. *Information Sciences*, 173, 281–304.

Keogh, E., S. Chu, D. Hart, and M. Pazzani. 2001. An online algorithm for segmenting time series. *Proceedings of the 2001 IEEE International Conference on Data Mining*, IEEE Press, 289–296.

Oliver, J.J., R.A. Bexter, and C.S. Wallace. 1998. Minimum message length segmentation. *Proceedings of the Second Pacific-Asia Conference on Knowledge Discovery and Data Mining*, IEEE Press, 222–233.

Pedrycz, W., J.C. Bezdek, T.J. Hathaway, and G.W. Rogers. 1998. Two nonparametric models for fusing heterogeneous fuzzy data. *IEEE Trans. on Fuzzy Systems*, 6, 411–425.

Wang, Z.J. and P. Willett. 2004. Joint segmentation and classification of time series using class-specific features. *IEEE Transactions on Systems, Man. and Cybernetics, Part B: Cybernetics*, 34, 2, 1056–1067.

10

From Models to Granular Models

The use of information granularity is regarded as an essential design asset whose optimal allocation helps in augmenting existing models. In this chapter, we discuss several development scenarios and classes of numeric models in which the allocation of information granularity provides tangible benefits in order to better capture the experimental data, form granular conclusions, or offer a more detailed insight into the parameters of the original models.

10.1 Knowledge Transfer in System Modeling

In a nutshell, knowledge transfer is about forming ways that an existing source of knowledge (namely, an existing model) can be used in the presence of new, very limited, experimental evidence. In virtue of the nature of the problem at hand (a situation encountered quite commonly, e.g., in project cost estimation), new data could be very limited and this scarcity of data makes it insufficient to construct a new model. At the same time, the new data originate from a similar (but not the same) phenomenon (process) for which the original model has been constructed so the existing model, even though it could be applied, has to be treated with a certain level of reservation. Such situations can be encountered, for example, in software engineering where in spite of the existing similarities, each project, process, or product exhibits its own unique characteristics. Taking this into consideration, the existing model is generalized (abstracted) by forming its *granular* counterpart—a granular model where its parameters are regarded as information granules rather than numeric entities, namely, their nonnumeric (granular) version is formed based on the values of the numeric parameters present in the original model. The results produced by the granular model are also granular and in this manner they become reflective of the differences existing between the current phenomenon and the process for which the previous model has been formed.

In modeling systems, processes, and phenomena, we can regard a resulting model as an essential source of knowledge. Once it has been constructed on the basis of some usually quite large experimental data **D**, this source of knowledge can be used afterward for a variety of prediction, control, and description tasks thus contributing to the better understanding of the system. The quality of the model and usefulness depend upon the nature of the new scenarios in which the model is used. In particular, prior to its use it becomes essential to assess how these new situations are different from those manifested by the data used to construct the model. In many complex problems of planning and cost estimation of software projects, each scenario is quite different. The sources of knowledge (models) formed so far could be useful but must be treated with caution when being applied to new situations. Similarly, in analysis of financial data, there are a significant number of historical data but a limited collection of current data about possible health of financial institutions including records of their failures. Having models of failure based on the past data could be useful for the current data, but the results produced by such models require some interpretation.

Let us consider that for the current problem at hand we are provided with a very limited data set—some new experimental evidence **D**′. Given this small data, two possible scenarios could be envisioned:

(a) We can attempt to construct a new model based on the data **D**′. As the current data set is very limited, designing a new model does not look quite feasible: it is very likely that the model cannot be constructed at all, or even if formed, the resulting construct could be of low quality.

(b) We would like to rely on the existing model which although it deals with not the same situation, it has been formed on a large and quite representative body of experimental evidence. We may take advantage of the experience accumulated so far and augment it in a certain sense so that it becomes adjusted to the current albeit quite limited data. In doing this, we fully acknowledge that the existing source of knowledge has to be taken with a big grain of salt and the outcomes of the model have to be reflective of partial relevance of the model in the current situation. We quantify this effect by making the parameters of the model granular (i.e., more abstract and general) so that one can build the model around the conceptual skeleton provided so far. In this case, viewing the model obtained so far as a sound source of knowledge, we are concerned with the concept of an effective knowledge transfer. The knowledge transfer (which, in essence, is represented by some model denoted here by N) manifests in the formation of a more abstract version of the original model—a so-called granular model, G(N), where the granular nature of the model associates with the augmentation (abstraction) of the original model N being realized in the presence of new data.

The process of knowledge transfer is intuitively appealing and becomes visible in many endeavors. As a compelling example, let us consider models of quantitative software engineering (Pedrycz, Russo, and Succi, 2012).

We build models of processes and qualities of software. In software cost estimation, project planning, quality assessment, to name the main phases of the overall development process, we come up with some models whose construction heavily relies on collected experimental data. In several cases, fuzzy sets are used as a vehicle to capture a lack of detailed numeric data when dealing with software cost estimation.

Each software project is unique so the model designed on the basis of the previous data might not be completely relevant, however, at the same time, it cannot be altogether neglected. Building a model for this specific process or software quality might not be feasible—simply, one might have a very limited data set, especially in the case of an initial phase of the project or when there have not been substantial efforts to systematically collect data. In light of these reasons, we encounter knowledge transfer—here the available model is viewed only as an initial construct that requires more revising/adjustments.

In general, the essence of the process of knowledge transfer is illustrated in Figure 10.1. The original model, call it N, built on the basis of **D** is now *abstracted* through its granulation, yielding a granular version $G(N)$. This occurs when dealing with a new data **D′**. The granular model becomes more in rapport with the environment currently encountered. Furthermore, the level of granularity is regarded here to be an important design asset whose efficient or optimal allocation helps in an effective usage of knowledge already acquired on the basis of **D**. The allocation itself is regarded as an optimization vehicle to make the model more in rapport with the reality.

The generalization of the effect of knowledge transfer can be discussed in the case of "p" different sources of knowledge—models built on the basis of $\mathbf{D}_1, \mathbf{D}_2, \ldots, \mathbf{D}_p$, that is, $N_1, N_2, \ldots, N_p$. We are interested in determining such N_{i0}, which is abstracted (granulated) in the most efficient way. In other words, N_{i0} is the one for which $G(N_{i0})$ leads to the best representation (quantified by

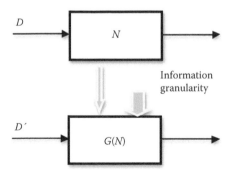

FIGURE 10.1
From model (N) to its granular counterpart ($G(N)$) being a result of the realization of knowledge transfer.

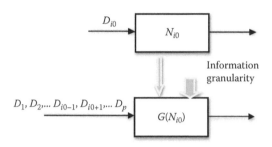

FIGURE 10.2
Formation of the best granular model among a family of locally constructed models N_1, N_2, ..., N_p.

means of some objective function) among all models available. Denoting this objective function of interest to us by Q, the problem is formulated as an optimization task of the form

$$i_0 = \arg\min{}_{i=1,2,...,p} Q(G(N_i)) \tag{10.1}$$

Again as before, a certain level of information granularity becomes available to form a granular version of the original model (refer to Figure 10.2 highlighting the very concept of knowledge transfer).

In what follows, we investigate the problem of knowledge transfer for a class of fuzzy logic networks.

10.2 Fuzzy Logic Networks—Architectural Considerations

The constructs of computational intelligence bringing together ideas of neurocomputing and fuzzy sets offer a great deal of synergy of resulting constructs. In what follows, we introduce a concept of fuzzy logic networks, elaborate on the underlying architecture as well as their interpretability, and look at some related design practices.

10.2.1 Realization of a Fuzzy Logic Mapping

Fuzzy sets and information granules, in general, offer a structural backbone of fuzzy logic networks. The crux of the concept is displayed in Figure 10.3. Information granules A_1, A_2, ... A_c are formed in the feature (input) space and output space. The information granules in the input space, B_1, B_2, ..., B_m are *logically* associated with the information granules positioned in the output space in the sense that for each of them a degree of activation is a logic function (logic mapping). All the information granules can be realized

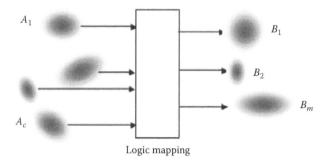

Logic mapping

FIGURE 10.3
An overall scheme of logic mapping between information granules—fuzzy sets in the input and output space and realized in the form of fuzzy logic network.

by exploiting several commonly encountered formalisms of information granulation including, for example, intervals (sets), fuzzy sets, and rough sets. In this study, we focus on the implementation of information granules in the form of fuzzy sets. When it comes to their detailed construction, we can resort to some well-known techniques of information granulation such as fuzzy clustering, for example.

As information granules are abstract, holistic descriptors of the existing data, all relationships among them can be conveniently captured in the form of logic dependencies, which in the sequel, can be realized in the form of a fuzzy logic neural network. The crux of such networks stems from the fact that they describe (and quantify) these relationships in the form of logic expressions, generalizing a logic characterization of dependencies present in Boolean expressions.

The flexibility and adaptive capabilities of the logic mapping is offered through the use of the collection of logic neurons (fuzzy neurons) whose connections are optimized during the design process of the model. Alluding to Figure 10.3, the relationship between the output of the network—an activation level of information granule B_j is treated as a truth value of the logic expression of the input information granules $A_1, A_2, \ldots, A$, that is, $B_j = L(A_1, A_2, \ldots A_c, \mathbf{W})$ with $\mathbf{W}$ being a vector of adjustable parameters of the logic expression (L).

We start by looking at the functional components of the network—logic neurons.

10.2.2 Main Categories of Aggregative Fuzzy Neurons: AND and OR Neurons

Logic neurons come with a clearly defined semantics of its underlying logic expression and are equipped with significant parametric flexibility necessary to facilitate substantial learning abilities. Formally, a logic neuron realizes a logic mapping from $[0,1]^n$ to $[0,1]$. Two main classes of the processing units are identified:

OR neuron: This neuron realizes an *and* logic aggregation of inputs $x = [x_1\ x_2\ ...\ x_n]$ with the corresponding connections (weights) $w = [w_1\ w_2\ ...\ w_n]$ and then summarizes the partial results in an *or*-wise manner (hence the name of the neuron). The concise notation underlines this flow of computing, $y = OR(x; w)$ while the realization of the logic operations gives rise to the expression (commonly referring to it as an s-t combination or s-t aggregation of the inputs and the corresponding connections)

$$y = \mathop{S}_{i=1}^{n}(w_i t x_i) \tag{10.2}$$

Recall that t-norms and t-conorms (s-norms) are the generic models of logic operators used in fuzzy sets. Some commonly encountered examples are the minimum and product (t-norms) and the minimum operator and the probabilistic sum (t-conorms). Bearing in mind the interpretation of the logic connectives (t-norms and t-conorms), the OR neuron realizes the following logic expression being viewed as an underlying logic description of the processing of the input signals

$$(x_1\ and\ w_1)\ or\ (x_2\ and\ w_2)\ or\ ...\ or\ (x_n\ and\ w_n) \tag{10.3}$$

Apparently the inputs are logically *weighted* by the values of the connections before producing the final result. In other words, we can treat "y" as a certain truth value of the above statement where the truth values of the inputs are affected by the corresponding weights. Noticeably, lower values of w_i discount the impact of the corresponding inputs; higher values of the connections (especially those positioned close to 1) do not affect the original truth values of the inputs resulting in the logic formula. In limit, if all connections w_i, $i = 1, 2, ..., n$ are set to 1 then the neuron produces a plain *or*-combination of the inputs, $y = x_1\ or\ x_2\ or\ ...\ or\ x_n$. The values of the connections set to zero eliminate the corresponding inputs. Computationally, the OR neuron exhibits nonlinear characteristics (that is inherently implied by the use of the t-norms and t-conorms (which are evidently nonlinear mappings). The connections of the neuron contribute to its adaptive character; the changes in their values form the crux of the parametric learning.

AND neuron: The neurons in the category, described as $y = AND(x; w)$ with x and w being defined as in case of the OR neuron, are governed by the expression

$$y = \mathop{T}_{i=1}^{n}(w_i s x_i) \tag{10.4}$$

Here the *or* and *and* connectives are used in a reversed order: first the inputs are combined with the use of the t-conorm (s-norm) and the partial results produced in this way are aggregated *and*-wise. Higher values of the connections reduce impact of the corresponding inputs. In limit $w_i = 1$ eliminates the relevance of x_i. With all connections w_i set to 0, the output of the AND neuron is just an *and* aggregation of the inputs

$$y = x_1 \ and \ x_2 \ and \ ... \ and \ x_n \tag{10.5}$$

Let us conclude that the neurons are highly nonlinear processing units whose nonlinear mapping depends upon the specific realizations of the logic connectives. They also come with potential plasticity whose usage becomes critical when learning the networks including such neurons.

At this point, it is worth contrasting these two categories of logic neurons with *standard* neurons we encounter in neurocomputing. The typical construct there comes in the form of the weighted sum of the inputs $x_1, x_2, ..., x_n$ with the corresponding connections (weights) $w_1, w_2, ..., w_n$ being followed by a nonlinear (usually monotonically increasing) function that reads as follows:

$$y = g(w^T x + \tau) = g\left(\sum_{i=1}^{n} w_i x_i + \tau \right) \tag{10.6}$$

where **w** is a vector of connections, τ is a constant term (bias), and "g" denotes some monotonically nondecreasing nonlinear mapping.

While some superficial and quite loose analogy between these processing units and logic neurons could be derived, one has to be cognizant that these neurons do not come with any underlying logic fabric and hence cannot be easily and immediately interpreted.

Let us make two observations about the architectural and functional facets of the logic neurons we have introduced so far.

Incorporation of the bias term (bias) in the fuzzy logic neurons. In analogy to the standard constructs of a generic neuron as presented above, we could also consider a bias term, denoted by $w_0 \in [0, 1]$, which enters the processing formula of the fuzzy neuron in the following manner:

$$\text{for the OR neuron,} \qquad y = \mathop{S}_{i=1}^{n} (w_i t x_i) s w_0 \tag{10.7}$$

$$\text{for the AND neuron,} \qquad y = \mathop{T}_{i=1}^{n} (w_i s x_i) t w_0 \tag{10.8}$$

We can offer some useful interpretation of the bias by treating it as some nonzero initial truth value associated with the logic expression of the neuron. For the OR neuron, it means that the output does not reach values lower than the assumed threshold. For the AND neuron equipped with some bias, we conclude that its output cannot exceed the value assumed by the bias. The question whether the bias is essential in the construct of the logic neurons cannot be fully answered in advance. Instead, we may include it into the structure of the neuron and carry out learning. Once its value has been obtained, its relevance could be established considering the specific value, which has been produced during the learning. It may well be that the optimized value of the bias is close to zero for the OR neuron or close to one in the case of the AND neuron which indicates that it could be eliminated without exhibiting any substantial impact on the performance of the neuron.

Dealing with inhibitory character of input information. Owing to the monotonicity of the t-norms and t-conorms, the computing realized by the neurons exhibits an excitatory character. This means that higher values of the inputs (x_i) contribute to the increase in the values of the output of the neuron. The inhibitory nature of computing realized by standard neurons by using negative values of the connections or the inputs is not available here as the truth values (membership grades) in fuzzy sets are confined to the unit interval. This is the same mechanism being encountered in a two-valued (Boolean) logic. The inhibitory nature of processing developed here can be accomplished by considering the complement of the original input that is $1-x_i$. Hence, when the values of x_i increase, the associated values of the complement decrease and subsequently in this configuration we could effectively treat such an input as exhibiting an inhibitory character.

10.2.3 An Architecture of the Fuzzy Logic Networks

The logic neurons can serve as building blocks of more comprehensive and functionally appealing architectures. The diversity of the topologies one can construct with the aid of the proposed neurons is surprisingly high. This architectural multiplicity is important from the application point of view as we can fully reflect the nature of the problem in a flexible manner. It is essential to capture the problem in a logic format and then set up the logic skeleton—a *conceptual* blueprint of the network (by forming the topology of the model and finally refine it parametrically through a thorough optimization of the connections). Throughout the entire development process we are positioned quite comfortably by monitoring the optimization of the network as well as interpreting its semantics.

The typical logic network that is at the center of logic processing originates from the two-valued logic and comes in the form of the fundamental Shannon theorem of decomposition of Boolean functions. Let us recall that any Boolean function $\{0,1\}^n \rightarrow \{0,1\}$ can be represented as a logic sum of its corresponding miniterms or a logic product of maxterms. By a minterm of "n" logic variables $x_1, x_2, ..., x_n$, we mean a logic product involving all these variables coming either in direct or complemented form. Having "n" variables we end up with 2^n minterms starting from the one involving all complemented variables and ending up at the logic product with all direct variables. Likewise by a maxterm, we mean a logic sum of all variables or their complements. Now, in virtue of the decomposition theorem, we note that the first representation scheme involves a two-layer network where the first layer consists of AND gates whose outputs are combined in a single OR gate. The converse topology occurs for the second decomposition mode: there is a single layer of OR gates followed by a single AND gate aggregating *or*-wise all partial results.

The proposed network (referred here as a logic processor) generalizes this concept as shown in Figure 10.4. The OR-AND mode of the logic processor comes with the two types of aggregative neurons being swapped between the layers. Here the first (hidden) layer is composed of the OR neuron and is followed by the output realized by means of the AND neuron. The inputs and outputs are the levels of activation of information granules expressed in the input and output spaces.

The logic neurons generalize digital gates by bringing essential learning capabilities and expanding the construct from its Boolean version to the multivalued alternative. The design of the network (i.e., any fuzzy function) is realized through learning. If we confine ourselves to Boolean {0,1} values, the network's learning becomes an alternative to a standard digital design, especially a minimization of logic functions. The logic processor translates into a composite logic statement (for the time being we skip the

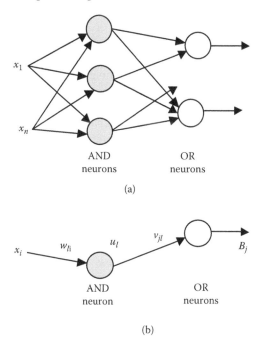

FIGURE 10.4

A topology of the logic processor (LP) in its AND-OR mode of realization (a) and the detailed notation (see the text).

connections of the neurons to emphasize the underlying logic content of the statement)

$$\text{truth value } (B_j) = (u_1 \; \textit{and} \; v_{j1}) \; \textit{or} \; (u_2 \; \textit{and} \; v_{j2}) \; \textit{or} \; ... \; \textit{or} \; (u_h \; \textit{and} \; v_{jh})$$

with

$$u_j = (w_{l1} \; \textit{or} \; x_1) \; \text{and} \; (w_{l2} \; \textit{or} \; x_2) \; \text{and} \; ... \; \text{and} \; (w_{ln} \; \textit{or} \; x_n)$$

where the truth value of B_j can be also regarded as a level of *satisfaction* (activation) of the information granule B_j in the output space. The above logic statements are quantified with the use of the connections of the logic processor. Given the number of inputs and the number of outputs equal to "n" and "m," respectively, the logic processor generates a mapping from $[0,1]^n$ to $[0,1]^m$ thus forming a collection of "m" n-input fuzzy functions.

10.2.4 Allocation of Information Granularity

Intuitively, as the connections are granular (interval-valued), the output produced by the network is also of interval-valued nature. Ideally, one would anticipate that the outputs of the granular network should include the original data **D'**. Consider $x_k \in \textbf{D}'$ with the cardinality of **D'** equal to N'. After the transformation of x_k by the elements of A yields the vector $z_k \in [0,1]^n$. This means that each of the outputs of the granular neural network comes in the form of the interval $Y_{kj} = [y_{j-}, y_{j+}] = G(N(z_k))_j$, j = 1, 2,..., m. Furthermore, through the transformation realized by the elements of B, f_k translates into **target$_k$** $= B(f_k) \in [0,1]^m$. As discussed in some previous cases, the quality of the granular network can be assessed by counting how many times the inclusion relationship target$_{kj} \in G(N(x_k))_j$ is satisfied. In other words, the performance index is expressed in the following form:

$$\kappa = \frac{\sum\limits_{j=1}^{m} \sum\limits_{k=1}^{N'} \{\text{card}((k, j) \, | \, \text{target}_{kj} \in G(N(z_k)_j)\}}{N'*m} \qquad (10.9)$$

Ideally, we could expect that this ratio is equal to 1. Of course κ becomes a nondecreasing function of ε, $\kappa(\varepsilon)$ so less specific information granules (higher values of ε) produce better coverage of the data but at an expense of the results obtained being less specific. Note also that the values of κ depend upon the predetermined level of ε, emphasized here by the notation $\kappa(\varepsilon)$. Here a monotonicity property is satisfied, namely, $\kappa(\varepsilon)$ is a nondecreasing function. The global assessment of the quality of the granular model is expressed as $\kappa = \int_0^1 \kappa(\varepsilon) \, d\varepsilon$. Along with the coverage criterion, we can look at the quality of the information granule of the output formed by the granular logic network, that is a length L of the interval $L(G(N(\mathbf{x}_k)))$ or its average value,

$$L = \frac{1}{M} \sum_{k=1}^{M} L(G(N(\mathbf{x}_k))) \tag{10.10}$$

Note that the criteria expressed by Equations (10.9) and (10.10) are in conflict: while high values of (10.9) are preferred, lower values of (10.10) are observed. If the two criteria are going to be considered at the same time, then the formation of a Pareto front is a way to proceed in the optimization process. As before, the protocols of allocation of information granularity are of relevance here:

P$_1$: Uniform allocation of information granularity. This process is the simplest one and in essence does not call for any optimization. All weights (connections) are treated in the same way and become replaced by the same interval (the length of the interval is the same for all connections).

P$_2$: Uniform allocation of information granularity with asymmetric position of intervals around the original connections of the network.

P$_3$: Nonuniform allocation of information granularity with symmetrically distributed intervals of information granules.

P$_4$: Nonuniform allocation of information granularity with asymmetrically distributed intervals of information granules.

P$_5$: A random allocation of granularity. This serves as a reference protocol.

In all these protocols, we ensure that the allocated information granularity meets the constraint of the total granularity, that is, εH where H stands for a number of all the connections of the network.

10.3 Granular Logic Descriptors

Local sources of knowledge structured in the form of logic descriptors—constructs of fuzzy logic, are arranged together (structured) in the form of a global model coming as a high-level granular logic descriptor. The inherent granularity of the global descriptor of this nature arises as a manifestation of the diversity of the locally available descriptors. The granular descriptor can be expressed with the aid of any of the formal models of information granules including sets, fuzzy sets, rough sets, probabilistic granules, and others. The architectural essence of the granular descriptor, which supports a quantification of the variability among the sources of knowledge, is realized through an optimal allocation of information granularity. Information granularity is treated as an important design asset and its allocation throughout the parameters of the logic descriptors helps quantify the diversity of individual sources of knowledge. Various protocols of allocation of information granularity along with an overall quantification of their effectiveness are discussed along with their numeric characterization.

Quite often one encounters modeling scenarios where a number of *local* sources of knowledge become available and need to be used *en block* in further processing to arrive at a holistic view of the phenomenon under discussion. This diversity of these sources has to be taken into account when constructing piece knowledge of a *global* nature. For instance, considering that the local sources are descriptors of some decision-making processes realized by humans (and in this way exhibiting a quite local character confined to a single individual), we are interested in retaining and quantifying the diversity of the local sources of knowledge when arriving at the model formed at the higher level of abstraction. Each decision-maker comes with a local model of decision—by ranking possible decision actions. A collective (group) decision-making naturally gives rise to some ranking agreeable to the group with an indication as to the diversity of the preferences and opinions being expressed within the group. Likewise, the effect of hesitancy resulting because of the diversity in the points of view is captured. A similar effect is observed when fusing classifiers. In this situation, each classifier is a local source of knowledge being reflective of realization of some local views at the classification problem (classification data). Taking these classifiers together offers an interesting alternative of carrying out classification results at the global, more general level. The classifiers may produce different results. They have to be reconciled by invoking some mechanisms of consensus building. The final outcome should be reflective of the existing diversity offering an important overview of the classification pursuits completed so far and, if necessary, produce some guidelines as to the enhancements of the local sources of knowledge (classifiers).

In contrast to the plethora of aggregation/consensus building approaches encountered in the literature where such mechanisms are typically invoked and return a single numeric realization positioned at the same level of abstraction as the results locally available are numeric, our position is that in order to properly cope with the construct arising at the global level, it has to be formed as more abstract (general) than the individual locally available characterizations of available local knowledge. When the term of generality comes into the picture and becomes contrasted with the numeric results, a notion of information granularity and subsequently information granules emerges as an important alternative worth considering. Information granules are intrinsic generalizations of numeric information (outcomes). Building a conceptually richer granular descriptor at the higher level of abstraction can dwell upon a suitable use of existing mechanisms of information granules.

The local logic descriptors are reflective of the locally available knowledge. The global descriptor, referred to as a *granular* logic descriptor, organizes the local constructs and quantifies their diversity. The notion of information granularity arises as an important design mechanism. The quantification of the existing diversity translates into a certain level of information granularity and an ensuing way of allocation of information granularity across the parameters of the global knowledge thus making it granular. It is shown that the process can be sought as an optimization of allocation of information granularity. The level of information granularity is viewed as an important design asset delivering a great deal of flexibility to the resulting granular model.

10.3.1 Logic Descriptors: Quantified *and* and *or* Logic Structures

We define a logic descriptor as a certain logic expression coming in a conjunctive or disjunctive form. The *ii*-th logic descriptors involving "n" variables and coming in a conjunctive form read as follows:

$$L_{ii}: y[ii] = (w_1[ii] \text{ or } x_1) \text{ and } (w_2[ii] \text{ or } x_2) \text{ and } .. \text{ and } (w_n[ii] \text{ or } x_n) \quad (10.11)$$

where $\mathbf{w}[ii] = [w_1[ii] \ w_2[ii] \ ... \ w_n[ii]]$ is a vector of weights (connections, parameters) and $\mathbf{x} = [x_1 \ x_2 ... \ x_n]$. The descriptor L_{ii} is regarded as linguistically quantified expression: it is an aggregation of inputs with the corresponding weights capturing degrees of satisfaction of the individual logic variables combined *and*-wise, namely, x_1 *and* x_2 *and* ... *and* x_n. In terms of computing, the flow of processing is schematically shown in Figure 10.5.

The corresponding connections (weights) are used to quantify a level of contribution of the individual variables to the output. The above aggregation is implemented with the use of t-norms and t-conorms meaning that the degree of satisfaction of L_{ii} is computed in the form of a t-s composition

$$L_{ii}: y[ii] = (w_1[ii] s \ x_1) \ t \ (w_2[ii] s \ x_2) \ t \ .. \ t \ (w_n[ii] s \ x_n) \quad (10.12)$$

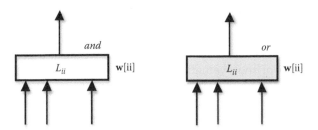

FIGURE 10.5
A schematic input–output representation of logic descriptors: (a) *and* logic descriptor, (b) *or* logic descriptor.

where "t" is a certain t-norm and "s" is a t-conorm. On the other hand, the *ii*-th logic descriptor of disjunctive form comes as follows:

$$L_{ii}: y[ii] = (w_1[ii] \text{ and } x_1) \text{ or } (w_2[ii] \text{ and } x_2) \text{ or } .. \text{ or } (w_n[ii] \text{ and } x_n) \quad (10.13)$$

where again, as in the case of the conjunctive logic expression Equation (10.12), we use some t-norms and t-conorms to aggregate the inputs. This results in the detailed formula involving t-norms and t-conorms

$$L_{ii}: y[ii] = (w_1[ii] t \, x_1) \, s \, (w_2[ii] t \, x_2) \, s.. \, s \, (w_n[ii] t \, x_n) \quad (10.14)$$

ii = 1, 2,..., c. As in the case of the *and* logic descriptors, this logic descriptor is illustrated in Figure 10.5. In essence, each logic descriptor $L_1, L_2, ..., L_c$ is a logic mapping from $[0,1]^n$ to $[0,1]$.

Each individual logic descriptor captures a locally available source of knowledge. It is worth noting that such sources could exhibit a great deal of diversity.

As an illustrative yet practically relevant example, we consider a decision-making scenario in which we encounter a collection of criteria with regard to which the suitability of a certain alternative is evaluated. Each criterion contributes to this assessment process. When dealing with a collection of decision makers, each of them has established his own evaluation model. In a nutshell, the relationship between the levels of satisfaction of the criteria and the resulting satisfaction level of the alternative is of logic nature where we aggregate these levels *and*- or *or*-wise (obviously taking into account the corresponding relevance levels). Either of these aggregations schemes formalizes in the form of dependency expressed by Equations (10.11) or (10.13), that is,

satisfaction level = satisfaction (criterion₁) *and* satisfaction (criterion₂)
and... *and* satisfaction (criterionₙ)

or

satisfaction level = satisfaction (criterion₁) *or* satisfaction (criterion₂)
or... *or* satisfaction (criterionₙ)

In the case of a group of decision-makers, it becomes apparent that each participant might have some opinion as to the relevance of the criteria, which becomes reflective in the obtained values of the weights.

To account for the diversity and quantify it, we form a general, namely, a granular form of the evaluation model. The granular weights obtained there become reflective of the existing opinions whereas the admitted level of information granularity expresses the extent to which the opinions can be kept different. Furthermore, once the optimal allocation of information granularity has been completed, the size of the information granules of the weights formed in this manner can serve as indicators of consistency of importance of the criteria across the group of decision-makers. The most

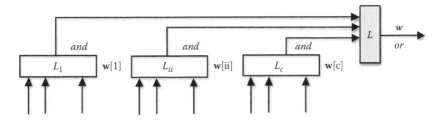

FIGURE 10.6
A schematic input–output representation of a logic processor.

specific (narrow) interval identifies the criterion for which the opinions of experts are highly consistent.

In addition to the plain (generic) disjunctive and conjunctive form of descriptors, one can consider more comprehensive and functionally appealing logic architectures in which these descriptors are put together forming a so-called logic processor. The typical logic network that is at the center of logic processing originates from the two-valued logic and comes in the form of the fundamental Shannon theorem of decomposition of Boolean functions. Now in light of the decomposition theorem, we note that the first representation scheme involves a two-layer architecture where the first layer consists of conjunctive descriptors while the second one is formed by means of the disjunctive descriptors presented above. The schematic view of the processing is illustrated in Figure 10.6.

10.3.2 The Development of Granular Logic: A Holistic and Unified View of a Collection of Logic Descriptors

We have started with a collection of logic descriptors describing some *local* logic; pieces of knowledge are arranged together to form a *global* description of available knowledge. Given the available diversity (which we intend to represent and quantify) of the local sources of knowledge, it is legitimate to contemplate that a holistic logic description emerges at the higher level of abstraction, which can be referred to as a *granular* logic descriptor. Information granularity inherent to this logic descriptor arises here as a result of an inherent diversity present within a family of individual descriptors. Furthermore, what is also important is an ability of the granular construct to reflect upon and quantify the existing diversity of logic descriptors. This unified and holistic view of the logic descriptors comes in the form of a certain granular abstraction of $\{L_1, L_2, \ldots L_c\}$, denoted here as $G(L)$ and schematically captured in Figure 10.7. The local logic descriptors could be either disjunctive or conjunctive in nature.

More formally, the global descriptors come in a conjunctive or disjunctive form, that is,

$$L: y = (W_1 s\ x_1)\ t\ (W_2\ s\ x_2)\ t \ldots t\ (W_n\ s\ x_n) \tag{10.15}$$

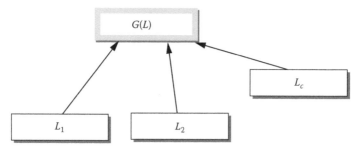

FIGURE 10.7
From local logic descriptors L_i to their global granular description $G(L)$.

or

$$L: y = (W_1 t \ x_1) \ s \ (W_2 t \ x_2) \ s.. \ s \ (W_n t \ x_n) \tag{10.16}$$

where W_j is a granular weight (connection) coming as a result of the abstraction of the family of local descriptors.

The two general modes of the realization of the granular logic descriptor are considered:

(a) Feedforward development

(b) Feedback-supported development

As this taxonomy stipulates, in the feedforward mode, one collects the results of the local logic descriptors and constructs the overall granular logic descriptor. Different logic descriptors can be selected to start with when forming a granular counterpart and different strategies can be employed to form such a construct, however, the original logic descriptors are not affected. In contrast to this feedforward mode, in the feedback type of development of granular descriptors, we witness an active role of the individual logic descriptors, which not only supply results to be used at the global level but also are subject to adjustments depending upon the findings formed at the higher level. In this sense, through such feedback all components are engaged in the overall process of establishing the global perspective.

Given the suite of the protocols, one can organize the feedforward construction of the global logic descriptor for a given protocol of allocation of information granularity P in the following manner,

For ii $= 1, 2, ..., c$

- Treat L_{ii} as a numeric blueprint of the global granular descriptor for the protocol P
- Optimize Q (treated either as a single or two-objective optimization task) and compute the AUC (whose value is obtained by covering the whole range of values of ε).

End

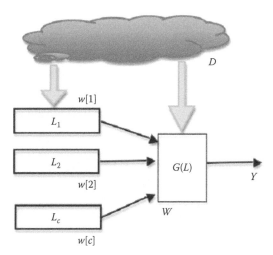

FIGURE 10.8
A schematic view of the design of a granular logic descriptor G(L).

The global descriptor is the one constructed on the basis of L_{ii} returning the highest value of its AUC. Let us stress that the optimization is carried out in the scenario determined by the data set D and the protocol P. Any changes to these two components will impact the resulting global granular descriptor. An overview of the formation of the global granular descriptor with a visualization of these two realized at the global level is illustrated in Figure 10.8. The data $x(k)$ belonging to D are used to determine the corresponding outputs of L_is (target$_k$) and then the resulting pairs of input–output data $(x(k), target_k)$ are used in the optimal allocation of information granularity to the weights of the granular logic descriptor.

10.3.3 The Development of the Granular Logic Descriptor in a Feedback Mode

The above scheme could be augmented by some feedback loop where we assess the quality (performance) of the individual logic descriptor vis-à-vis the global view being formed. This could help eliminate or reduce the impact of some logic descriptors that are quite distant from the rest of the local descriptors. For the realization of the feedback loop we consider, as before, the data set $D = \{(x(k)\}, k = 1, \ldots, N$. Likewise the granular descriptor being formed produces the granular output for every input $x(k)$ in D, that is, $Y_k = G(L(x(k)))$.

Then we assess the performance of each local descriptor L_i by counting the number of cases the result $L_i(x(k))$ is included in Y_k formed at the global level. More specifically, we compute the ratio $g_i = card\{x(k) \in D_i \mid L_i(x(k)) \subset Y_k\}/N_i$, namely, count the number of cases where the data D_i are covered by the interval Y_k. The number of input–output pairs in D_i is equal to N_i. The higher the value of this ratio, the higher the consistency of the descriptor L_i with the granular descriptor achieved at the global level. Using the above ratio we realize a feedback mechanism to achieve higher consistency of the local descriptors guided by the performance of the granular descriptor. The *i*-th logic descriptor is redesigned by minimizing the augmented performance index

$$Q = \sum_{target,x \in D_i} (L_i(x) - target)^2 + \zeta \sum_{x \in D} (L_i(x) - num(Y))^2 \qquad (10.17)$$

where $\xi_i = 1 - g_i$. The notation num(Y) stands for a numeric representative of the information granule Y generated by the granular logic descriptor; for instance, it could be taken as the center of the interval or a center of gravity of the fuzzy set. In a nutshell, it is an additive augmentation of the original performance index used to the optimization of L_i. The weight factor ξ_i is essential in invoking the redesign (optimization) of L_i in light of its consistency with $G(L)$. If $\xi_i = 1$, this means that the originally developed L_i is fully consistent with the granular descriptor so no further relearning of it is required. If the values of ξ_i are higher then more intensive learning becomes necessary (as the second component of Equation 10.17 gets higher). Once the local logic descriptors have been updated through the minimization of the sum of squared errors

$$V = \sum_{x \in D_i} (L_i(x) - target)^2 \qquad (10.18)$$

then they are used to develop of $G(L)$ and the process is repeated. The overall scheme of the successive development is displayed in Figure 10.9.

The feedback-oriented design of the granular descriptor is evidently iterative where each iteration involves the determination of the granular construct and the following redesign (update) of the local descriptor. More concisely, we can express the scheme as follows:

Start: construct L_i using locally available data D_i.
Iterate (iter = 1, 2...)

- Design the granular model $G(L)$
- Update logic descriptors L_i by minimizing Equation (10.17)

Until a certain stopping criterion has been satisfied.

The stopping criterion could be realized in different ways. In essence, we monitor the improvement of the granular descriptor and once the process stabilizes, it is terminated.

We note that the overall scheme is invoked in the context of some data *D* so a change in the data might result in a different granular logic descriptor.

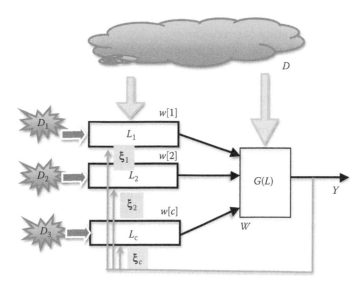

FIGURE 10.9
The design of the granular descriptor $G(L)$ including a feedback loop driving the development scheme and further refinements (relearning) of the local logic descriptors.

This points to the role of the data in the overall construct and underlines a need for its representativeness in describing the characteristics of the logic descriptors.

10.4 Granular Neural Networks

Most neural networks encountered in the literature are *numeric* constructs realizing a certain nonlinear mapping. A conceptually viable and practically useful generalization of numeric neural networks comes in the form of nonnumeric mappings realized by neural networks. In this case, we may refer to such networks as *granular* neural networks. The nonnumeric (granular) nature of the mapping arises because of the granular character of the connections. In this case any numeric input to such network produces a granular output of the network. There are several compelling reasons behind the realization of this type of neural network. First, by establishing granular outputs, one can effectively gauge the performance of the already constructed numeric neural network in the presence of training data. Second, when dealing with new data, the network forms granular outputs, which are instrumental in the quantification of the quality of the obtained result. For instance, in case of prediction, we are provided with a comprehensive forecasting outcome: instead of a single numeric result, an information granule is formed whose location and level of information granularity are highly descriptive of the quality of prediction.

The term *granular* pertains to the nature of the developed construct and is by no means confined to a certain specific type of the neural network. Instead, it concerns a general augmented neural architecture it builds upon. The proposed concept applies equally well to multilayer perceptrons (MLPs) or radial basis function neural networks (RBFNN) (Liu and Li, 2004; Wedge et al., 2006). It works well with neurofuzzy systems (Juang and Lin, 1999; Park et al., 2009; Ishibuchi, 1996; Buckley and Hayashi, 1994).

At this point, it is instructive to relate the proposed approach with what is known in the literature as interval neural networks, especially interval MLPs. Ishibuchi and Tanaka (1993) proposed a neural network with interval weights and interval biases and derived a learning algorithm supporting its development. The studies presented in Ishibuchi, Kwon, and Tanaka (1995), being a continuation of the previous work generalize numeric inputs of fuzzy neural networks to their fuzzy set-based counterparts. In Zhang et al. (2000) a granular neural network using backpropagation algorithm and fuzzy neural networks is used to handle numeric–linguistic data fusion providing a mechanism of knowledge discovery in numeric–linguistic databases. It is noticeable that in all these cases, the corresponding interval neural networks are built from scratch. Some other interesting developments were reported by Bortolan (1998), Zhang, Jin, and Tang (2008), de Weerdt, Chu, and Mulder (2009), and Pedrycz and Vukovich (2001). In spite of interesting concepts and design schemes, all of these constructs exhibit a clearly visible commonality: they are formed on the basis of some granular rather than numeric data and do not dwell upon the well-established learning schemes applied to numeric data and resulting in numeric constructs.

In contrast, in the proposed neural architecture, we are concerned with granular connections, in general, and this implies various formal models of information granules including sets, fuzzy sets, probabilities, and rough sets. From a computational perspective, intervals offer a computationally appealing alternative and while the underlying concepts are of general character and those are relevant to other formalisms of information granules, more detailed investigations will be focused on interval-valued (interval) neural networks. Furthermore given the focus of the study, the term information granule and interval are used interchangeably. The entire design process behind the proposed networks is outlined as follows. A starting point of the overall design process is a numeric network that has already been developed by means of one of the well-established learning strategies. Then, a data set (the same training data set or a new one) is used to construct a granular network, namely, form interval connections on the basis of the given network. In this sense, the resulting granular construct augments the topology of the existing neural architecture. The design process (i.e., the formation of information granules of the connections) is well articulated and translates into an

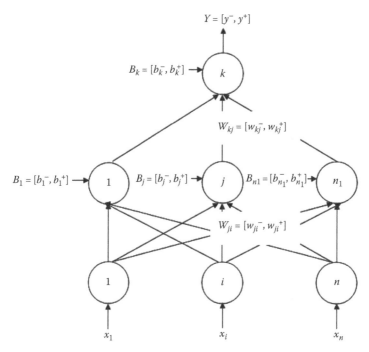

FIGURE 10.10
Architecture of a granular neural network.

optimization problem. Information granularity is regarded as an essential design asset and an allocation of granularity following some protocols leads to the optimization of some performance index gauging the quality of the resulting granular neural network.

The granular neural network under consideration, presented in Figure 10.10, comprises a single hidden layer consisting of n_1 neurons, and a single neuron located in the output layer. The features (inputs) to the network are organized in a vector form $\mathbf{x} = [x_1, x_2, ..., x_n]^T$. The weight (connection) connecting the i-th neuron in the input layer to j-th neuron in the hidden layer comes in the form of an interval and is denoted by W_{ji}, $W_{ji} = [w_{ji}^-, w_{ji}^+]$. The weights between the hidden layer and output layer are also interval valued. Each neuron comes with an interval bias. In virtue of the interval connections used in the network, for any numeric input, the result of processing becomes an interval, that is, $Y = [y^-, y^+]$.

As indicated, in our design, we proceed with the already constructed MLP (e.g., realized in terms of batch backpropagation, BP, learning method). The choice of the size of the hidden layer is decided upon during this design phase. The activation functions of the neurons in the hidden layer and the output layer are monotonically increasing functions.

As the outputs of the granular neural network are intervals (computed on the basis of the fundamental formulas provided in Chapter 2) while the targets coming from the experimental data are numeric, we have to define a suitable performance index (objective function), whose optimization (maximization or minimization) is realized through a suitable allocation (distribution) of information granularity.

The challenging yet highly important issue is how to construct interval-valued weights and biases of a network. The available information granularity (more specifically, its level of granularity), being treated as an important design asset, has to be carefully distributed among all the connections of the network so that the interval-valued output of the neural network covers (includes) the experimental datum. In what follows, we propose several protocols of allocation of information granularity and discuss two indexes whose optimization is realized through this allocation process.

10.4.1 Design Issues of the Granular Neural Networks

Given a level of information granularity ε assuming values in the unit interval, this level is allocated to the individual weights and biases. A way of building intervals around the original numeric values of the weights and biases can be referred to as granularity allocation. The allocation leading to the optimization of a given performance index refers to *optimal* information granularity allocation. The original weight or bias, denoted symbolically by w_{ji} and b_k, is made interval by forming some bounds around the original numeric values in the following way:

$$w_{ji}^- = w_{ji} - \varepsilon |w_{ji}| \quad b_k^- = b_k - \varepsilon |b_k| \tag{10.19}$$

$$w_{ji}^+ = w_{ji} + \varepsilon |w_{ji}| \quad b_k^+ = b_k + \varepsilon |b_k| \tag{10.20}$$

Note that for the parameter ε (or its refinements depending upon the use of the protocol of allocation of information granularity, that is, ε_- and ε_+ or ε_{ij}, etc.) used in the way shown above forms the bounds of the interval whose length is proportional to the original numeric value of the connection. The allocation of information granularity is realized by following the protocols discussed so far with the maximization of the coverage criterion.

Illustrative Example

Here we consider a two-variable sine wave, which has been used as regression benchmark data in numerous experiments reported for neural networks.

$$y = 0.8 \times \sin(x_1/4) \times \sin(x_2/2) \tag{10.21}$$

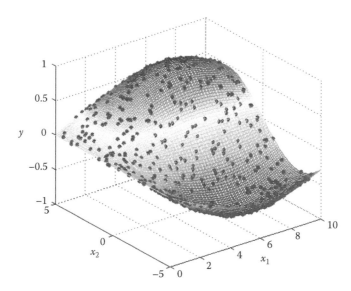

FIGURE 10.11
Nonlinear two-variable function along with training data.

where x_1 is in [0, 10] and x_2 is in [–5, 5]. The training data are made up of 480 randomly selected input–output pairs (following a uniform distribution over the input space). The function along with the superimposed training data is illustrated in Figure 10.11.

The performance of the constructed neural network quantified in terms of the mean squared error (MSE) is illustrated in Figure 10.12. Through a visual inspection, we choose the number of neurons in the hidden layer to be equal to eight.

The plots of the coverage regarded as a function of ε are included in Figure 10.13. The values of the AUC are computed over the unit interval, however the plot is shown only for a portion of the entire range of ε.

The corresponding values of the AUC on training data are: protocol P_1: AUC = 0.9755, protocol P_2: AUC = 0.9762, protocol P_3: AUC = 0.9864, protocol P_4: AUC = 0.9881, protocol P_5: AUC = 0.9754.

As shown in this figure, the value of AUC obtained for protocol P_4 is slightly higher than the one produced for P_3. However, both are higher than the ones produced by the other three protocols. This is not surprising as they reflect the increasing flexibility of the successive protocols of allocation of granularity and a better, more effective usage of information granularity.

Figure 10.14 displays the values of the fitness function reported in consecutive generations; it is noticeable that the convergence of the optimization process (increasing coverage) occurs in the first 25 generations.

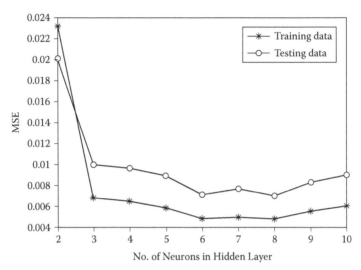

FIGURE 10.12
Performance index as a function of the number of neurons in the hidden layer.

The granular connections of the network are displayed in Figure 10.15 ($\varepsilon = 0.02$) and Figure 10.16 ($\varepsilon = 0.2$); both the lower and upper bounds of the intervals of the connections are visualized in the form of vertical bars.

The optimized coverage of this case is 0.9042. We use the same range for the y-axis on each plot of intervals of granularities. Refer to the enlarged plot

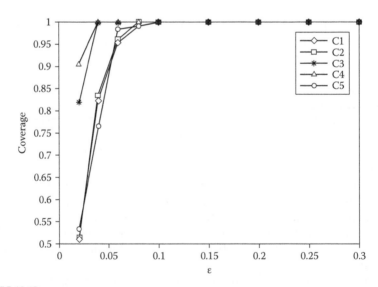

FIGURE 10.13
Coverage as a function of ε for the protocols of information granularity allocation used in the study.

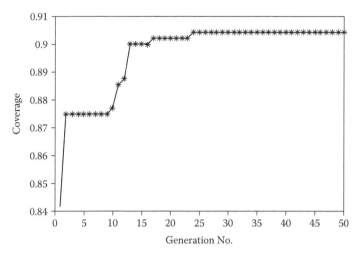

FIGURE 10.14
Performance of PSO expressed in terms of the fitness function obtained in consecutive genera-tions; $\varepsilon = 0.02$, protocol P_4.

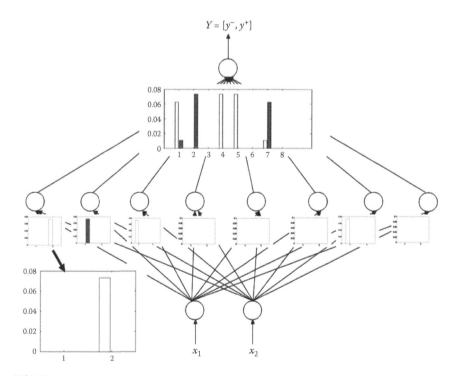

FIGURE 10.15
The optimized allocation of granularity levels realized by running protocol P_4 with $\varepsilon = 0.02$. The biases of the neurons are not shown.

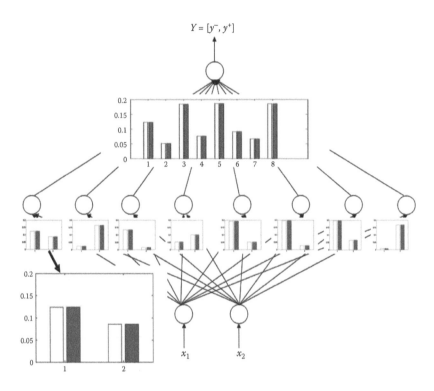

FIGURE 10.16
The optimized allocation of granularity of level 0.2 (ϵ) for P_4.

of one of the neurons positioned in the first layer for more detail. Since each interval is constructed by extending the numeric value to a lower bound and an upper bound separately, we actually form two subintervals for each connection. And in this way, we have to distribute the bounds of granularity to each connection. There are some connections, which have zero values for the levels of granularity, which means that these connections are effectively retained as numeric ones.

10.5 The Design of Granular Fuzzy Takagi–Sugeno Rule-Based Models: An Optimal Allocation of Information Granularity

10.5.1 General Observations

There have been many studies on the design, analysis, and implementations of Takagi–Sugeno (TS) fuzzy models and fuzzy models, in general. These models are considered to be one of the commonly encountered constructs of fuzzy modeling. In the area of analysis, we can refer to some recent studies presented in the literature (Jafarzadeh, Fadali, and Sonbol, 2011; Kim, Kim, and Lee, 2006; Lughofer, 2008). The dominant design trends are discussed in Juang and Hsieh (2012), Pal and Saha (2008), Li et al. (2012), and Kulkolj and Levi (2004).

When taking a close look at these constructs falling under the rubric of TS fuzzy models, we conclude that there are some remarkable commonalities among all of them. All of these fuzzy models are numeric constructs. For any input, computed is a corresponding numeric output. It is also evident that there are no ideal fuzzy models for which all data coincide with the results produced by the model. No matter what design practice one embraces and no matter how complex the architectures of the fuzzy models could be, there are no ideal models, which could fit all data (and produce a zero value of the performance index, that is, root mean squared error, RMSE).

To make the model perform more in rapport with reality, it would be beneficial to develop a fuzzy model producing outputs that are nonnumeric. In this way the model is made more abstract and reflective of the quality of the fuzzy model. The essence is that in order to quantify the quality of the fuzzy model, it becomes imperative to link the numeric results of the fuzzy model with the granular characterization of its results. Unfortunately, this has not happened in fuzzy modeling. A very loose analogy can be found in linear regression. Linear regression models are numeric, however, their numeric results are augmented by associated confidence intervals (and confidence curves), which offer a better insight into the quality of the model and the quality of a specific result. Similarly, confidence intervals of the parameters of the model produce a better view of the nature of the regression surface.

One of the first studies, which raised the issue of quantifying the quality of fuzzy models by stressing a genuine need to form a more abstract description of results of modeling and a more abstract characterization of the fuzzy model itself was discussed in Pedrycz (1990, 1998). The idea presented there can be treated as a precursor of what later resulted in the ideas of fuzzy modeling involving type-2 fuzzy sets or interval-valued fuzzy sets. Type-2 fuzzy models form an interesting and promising direction in fuzzy modeling, however, most of these models do not support a coherent methodological view in spite of useful experimental findings. This leads to an important observation that while dealing with fuzzy sets and fuzzy models, these models can be characterized at the higher level of abstraction articulated in terms of higher-order or higher-type information granules. The way of reasoning introduces a possibility of expressing the quality of fuzzy models by forming granular fuzzy models.

The quality of the model can be quantified by incorporating some granular constructs as discussed in so-called linguistic models (Pedrycz and Vasilakos, 1999). Another approach to fuzzy modeling in which information granularity plays a pivotal role uses shadowed sets. The rule-based fuzzy model involving shadowed sets in a collection of rules produces results of inference whose relevance is quantified with the three-valued membership assignment forming the crux of shadowed sets, namely, full membership, no membership, and *unknown* membership (modeled as a unit interval). Probabilistic sets offer another formalism to enrich the generic constructs of fuzzy sets (Hirota and Pedrycz, 1983; Czogala and Pedrycz, 1983).

Let us start with a simple observation, which motivates a general line of thought of this study. As the rule-based model consists of rules, an obvious question is about the performance of the overall model along with a distribution of errors within the input space. When plotting a distribution of error vis-à-vis a location of the input expressed with regard to the centers of the fuzzy sets forming conditions of the rules in the rule-based system, one notices that higher values of error are present for the inputs positioned in-between the modal values of the fuzzy sets present in the condition part of the rules.

By analogy with the commonly used form of the TS rule-based model composed of the rules

$$-\text{if } \mathbf{x} \text{ is } A_i \text{ then } y = f_i(\mathbf{x}, \mathbf{a}_i) \qquad (10.22)$$

we express the granular rule-based model as a collection of rules

$$-\text{if } \mathbf{x} \text{ is } G(A_i) \text{ then } y = f_i(\mathbf{x}, \mathbf{a}_i) \qquad (10.23)$$

where $G(A_i)$ denotes a granular generalization (abstraction) of the original fuzzy set of condition A_i. This granular abstraction can be realized in the form of an interval-valued fuzzy set (or type-2 fuzzy set, in general), rough set, shadowed set (as mentioned above), probabilistic set, and so forth. The conclusion part (f_i) includes a certain local function. In virtue of the granular nature of the condition part, the result (output) of the rule-based system is also of a granular nature, which we denote here by a capital letter Y, that is, $Y = f(\mathbf{x}, G(A_i), f_i, \mathbf{a}_i)$. As a matter of fact, the granularity of the result helps us make the model more realistic by including (covering) the experimental data within the bounds of the information granules.

10.5.2 Design of Takagi–Sugeno Fuzzy Models: Some General Views and Common Development Practices

As noted, the TS fuzzy models are composed of rules in the form Equation (10.22) where A_i is a multivariable information granule (fuzzy set) defined in the input space ($\mathbf{R}^n$) and f_i is a local function $\mathbf{R}^n \rightarrow \mathbf{R}$ equipped with some parameters $\mathbf{a}_i$. The processing carried out within the model consists of two steps: (a) determination of activation of the individual rules, which for any $\mathbf{x} \in \mathbf{R}^n$ returns the values $A_1(\mathbf{x}), A_2(\mathbf{x}), \ldots, A_c(\mathbf{x})$, which could be treated as activation levels of the corresponding rules, and (b) aggregation of outcomes of local models f_i weighted by the activation levels. The aggregation is typically realized as the following sum $y = \sum_{i=1}^{c} A_i(x) f_i(x)$. The advantages of this category of the models are apparent. The modularity of the models and their local character are one of the visible modeling assets: even a complex phenomenon can be easily modeled by far less complex (e.g., linear) and local relationships, admitting that their relevance is confined to some quite limited regions of the input space. The paradigm of *local* rather than *global* modeling is behind the principle of rule-based modeling. It contributes to the success of this form of system modeling.

The design process comprises two main phases (whose realization could exhibit some algorithmic diversity) that is reflective of the rule-based architecture of the model:

(a) Construction of information granules,
(b) Construction of local modes forming the conclusion parts of the rules.

With regard to the first phase, the common practice is to form information granules forming the condition part of the rules through fuzzy clustering. Fuzzy c-means along with its numerous variants is one among well-established and commonly used techniques to determine fuzzy sets (whose description is based on both the partition matrix and the prototypes being determined during the clustering process). Based on the knowledge of the prototypes $v_1, v_2, ..., v_c$, the associated membership functions $A_1, A_2, ..., A_c$ are expressed through a well-known formula

$$A_i(x) = \frac{1}{\sum_{j=1}^{c} \left(\frac{\|x - v_i\|}{\|x - v_j\|} \right)^{2/(m-1)}} \tag{10.24}$$

where $\| . \|$ denotes a distance function (for the generic version of the FCM we use a Euclidean distance or its weighted version). The fuzzification coefficient (m) assumes values greater than 1 and impacts the geometry of the clusters (shape of membership functions).

If the local models (f_i) associated with the respective rules are linear with respect to their parameters then an estimation of their optimal values is realized by solving a standard least square error (LSE) problem for which there is an analytical solution assuring a global minimum of the optimization problem. Let us stress that most of the design of these models is realized in a supervised mode meaning that for clustering and parameter estimation we use a collection of input–output data (x_k, target$_k$), $k = 1, 2, ..., N$. The standard modeling practices of using these data for training and testing purposes are exercised here as well.

10.5.3 Granular Fuzzy Clusters

In the general architecture of granular TS models, Equation (10.23), the original numeric membership functions A_i are replaced by their granular counterparts $G(A_i)$ built on the basis of the fuzzy sets occurring in the rules, Equation (10.22). In what follows we look at the formation of granular membership functions. The starting points of the entire construction are granular prototypes V_1, V_2, ..., V_c formed around the numeric counterparts. More specifically, for the purpose of this study (and to focus on the essence of the construct), we assume that V_i are interval-valued prototypes, namely, some sets defined in the space of sets over R^n, that is, V_i $P(R^n)$ where $P(.)$ stands for a family of sets (intervals).

We discuss a way of constructing the granular prototypes. Because of their nonnumeric nature, granular prototypes give rise to granular membership functions. Note that the original formulas for the membership functions developed in the generic FCM, the membership grades (functions) are determined on the basis of the distances between x and the (numeric) prototypes. Here the notion of distance has to be carefully revisited to properly account for the interval nature of the prototypes. While there are some well-known approaches to express the distance between granular constructs (i.e., a Hausdorff distance) (Moore, 1966), all of them return a single numeric value quantifying this distance. This view is somewhat limited as one could have expected a certain granular descriptor of the closeness. The simplest option here would be to establish some bounds of the values the distance could assume. We consider these extreme cases by looking at a single variable. Let us assume that for the j-th variable, the bounds of the granular prototype V_i form the interval $[v_{ij}^-, v_{ij}^+]$. For the j-th coordinate of x, x_j, we consider two situations:

(i) $x_j \notin [v_{ij}^-, v_{ij}^+]$ The bounds of the distance are taken by considering the pessimistic and optimistic scenario and computing the distances from the bounds of the interval, that is, $\min((x_j - v_{ij}^-)^2, (x_j - v_{ij}^+)^2)$ and $\max((x_j - v_{ij}^-)^2, (x_j - v_{ij}^+)^2)$.

(ii) $x_j \in [v_{ij}^-, v_{ij}^+]$ It is intuitive to accept that the distance is equal to zero (as x_j is included in this interval).

The distance being computed on the basis of all variables $||x-V_i||^2$ is determined coordinatewise by involving the two situations outlined above. The minimal distance obtained in this way is denoted by $d_{min}(x, V_i)$ while the maximal one is denoted by $d_{max}(x, V_i)$. More specifically we have,

$$d_{min}(x, V_i) = \sum_{j \in K} \min((x_j - v_{ij}^-)2, (x_j - v_{ij}^+)^2)$$

$$d_{max}(x, V_i) = \sum_{k \in K} \max((x_j - v_{ij}^-)2, (x_j - v_{ij}^+)^2)$$

(10.25)

where $K = \{j = 1,2, \ldots, n \mid x_j \notin [v_{ij}^-, v_{ij}^+]\}$. Having determined the distances, we compute the two expressions

$$w_1(\mathbf{x}) = \cfrac{1}{\displaystyle\sum_{j=1}^{c} \left(\cfrac{d_{min}(\mathbf{x}, \mathbf{V}_i)}{d_{min}(\mathbf{x}, \mathbf{V}_j)} \right)^{1/(m-1)}}$$

$$w_2(\mathbf{x}) = \cfrac{1}{\displaystyle\sum_{j=1}^{c} \left(\cfrac{d_{max}(\mathbf{x}, \mathbf{V}_i)}{d_{max}(\mathbf{x}, \mathbf{V}_j)} \right)^{1/(m-1)}}$$

(10.26)

Notice that these two formulas resemble the expression used to determine the membership grades in the FCM algorithm. In the same way as in the FCM, the weighted Euclidean distance is considered here, namely, $d_{min}(\mathbf{x}, \mathbf{V}_i) = \Sigma_{j \in K} min((x_j - v_{ij}^-)^2/\sigma_j^2, (x_j - v_{ij}^+)^2/\sigma_j^2)$ and $d_{max}(\mathbf{x}, \mathbf{V}_i) = \Sigma_{j \in K} max((x_j - v_{ij}^-)^2/\sigma_j^2, (x_j - v_{ij}^+)^2/\sigma_j^2)$ with σ_j being the standard deviation of the j-th variable. These two are used to calculate the lower and upper bounds of the interval-valued membership functions (induced by the granular prototypes). Again one has to proceed carefully with this construct. Let us start with a situation when $\mathbf{x}$ is not included in any of the granular prototypes. In this case we compute

$$u_i^-(\mathbf{x}) = min(w_1(\mathbf{x}), w_2(\mathbf{x}))$$

and

$$u_i^+(\mathbf{x}) = max(w_1(\mathbf{x}), w_2(\mathbf{x}))$$ (10.27)

so, in essence, we arrive at the granular (interval-valued) membership function $U_i(\mathbf{x}) = [u_i^-(\mathbf{x}), u_i^+(\mathbf{x})]$. If $\mathbf{x}$ belongs to $\mathbf{V}_i$ then apparently $u_i^-(\mathbf{x}) = u_i^+(\mathbf{x}) = 1$ (and this comes as a convincing assignment as $\mathbf{x}$ is within the bounds of the granular prototype). Obviously, in this case $u_j^-(x)$ as well as $u_j^+(x)$ for all indexes "j" different from "i" are equal to zero.

In the construct above, we assume that an overlap between any two granular prototypes $\mathbf{V}_i$ and $\mathbf{V}_j$ does not happen. This argument is quite convincing in light of the moderate or high dimensionality of the space in which the prototypes are defined. Furthermore, the likelihood of overlap gets higher if the granular prototypes get larger, however one can control their level of granularity and prevent this from happening (as a matter of fact, this constraint will be implemented in the design procedure discussed later on).

As an illustration, let us consider a one-dimensional case with three interval-valued prototypes $V_1 = [0.95\ 1.3]$, $V_2 = [2.7\ 2.9]$, and $V_3 = [4.3\ 4.45]$. The lower and upper bounds of the resulting granular membership functions are displayed in Figure 10.17.

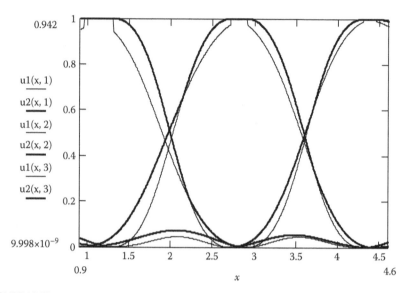

FIGURE 10.17
Example granular membership functions induced by granular prototypes (thick line denotes upper bounds of the membership functions).

As it might have been expected, the spread of the bounds of the granular membership function differ over the universe of discourse. As visualized in Figure 10.18, the regions where the highest differences in these membership grades are encountered are positioned in-between the prototypes. This could be expected, as those are the regions where contributions coming from

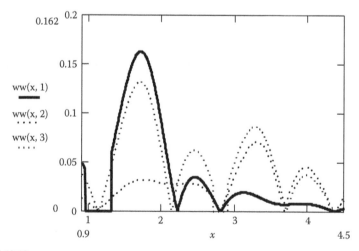

FIGURE 10.18
Differences between the bounds of the granular membership functions.

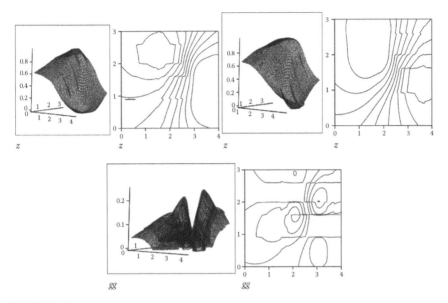

FIGURE 10.19
Upper and lower bounds of the granular membership functions of the first cluster and the difference between the bounds.

the prototypes are less substantial (and this effect becomes quantified by loose bounds of the membership values).

Again, following the illustrative example, we consider a two-dimensional case with the prototypes $V_1 = [1.5\ 1.8] \times [2.0\ 2.6]$ and $V_2 = [3.4\ 3.7] \times [0.9\ 1.6]$. The results are reported in a series of plots, Figure 10.19.

As a byproduct of the granular prototypes, one could think of them as offering an enhanced interpretation of clusters formed by the FCM. The numeric prototypes are not fully reflective of the nature of the underlying data. It might be that the two data sets could be composed of highly condensed clusters (first data sets) and clusters that are highly scattered (second data set), however they might have very similar prototypes. In the sequel, when we consider a new data point **x**, its manifestation in terms of the structure (clusters), namely, the membership grades (computed on the basis of Equation 10.24) are the same. It becomes apparent that the structure conveyed by the FCM offers a very preliminary description, which is not capable of discriminating between these two situations. In contrast, the granular prototypes help visualize the difference between the structures of these two data sets leading to the granular (interval-valued) membership values.

This presentation of granular membership functions links to a certain concept of granularity of prototypes discussed in Pedrycz and Bargiela (2012) where such augmented prototypes help describe a structure of data and identify patterns that are incompatible with the core structure.

10.5.4 Optimization of Granular Fuzzy Models

The performance index is used to gauge the quality of the effective usage of the available level of information granularity, which has been allocated to the individual prototypes. An intuitively appealing criterion one could consider here would be to check how well the original output data are covered by the granular output of the granular fuzzy model. More specifically, we compute a ratio of the number of cases $target_k$ is included in the interval $Y_k = f(x_k, G(A))$ to all data (N) and call it *coverage (ε)*

$$\text{coverage}(\varepsilon) = \frac{\text{number of cases where } target_k \subset Y_k}{N} \qquad (10.28)$$

In an ideal case, coverage(ε) achieves the value of 1. The better the allocation of granularity, the higher the value of (10.28). Note also that the coverage is a function of ε. As usual the optimization of the allocation of information granularity concerns the maximization of the coverage by distributing (allocating) $\varepsilon_1, \varepsilon_2, ..., \varepsilon_n$ so that the total balance of information granularity is retained, namely, $n\varepsilon = \sum_{i=1}^{n} \varepsilon_i$. Formally, we express the problem as follows

Max_ε coverage (ε) subject to $\varepsilon = [\varepsilon_1 \ \varepsilon_2 \ ... \ \varepsilon_n]$ such that $\varepsilon_1, \varepsilon_2, ... \ \varepsilon_n > 0$ and

$$n\varepsilon = \sum_{i=1}^{n} \varepsilon_i. \qquad (10.29)$$

As before, the optimization of allocation of information granularity is quantified with the use of the area under curve (AUC) computed in the following way

$$AUC = \frac{1}{\varepsilon_{max}} \int_{0}^{\varepsilon_{max}} \text{coverage}(\varepsilon) d\varepsilon \qquad (10.30)$$

where ε_{max} is the maximal value of the level of information granularity for which the granular prototypes still do not intersect.

10.6 Conclusions

We have shown that the design of granular models can be realized in a coherent way by exploiting the two principles of Granular Computing. The principle of justifiable granularity is more suitable to form granular outcomes of the model formed on the basis of a collection of individual sources of knowledge (local models) while the allocation of information granularity is more focused on the realization of the granular model to compensate for the discrepancies between the numeric model and the experimental numeric data. While these two principles were used separately when addressing the problems covered in this chapter, one can also envision situations where both of them can be used together in a single construct.

While the *granular* models were implemented here in terms of intervals, it was done for illustrative purposes—any other formalism of information granularity could be considered. A family of α-cuts (intervals) is one of the immediate alternatives to implement and construct fuzzy sets. The interactive mode of forming information granules or distributing information granularity realized in a feedback loop is helpful in the adjustments of the sources of knowledge (models) and arriving at the formation of consensus at the global level. The dynamics of this process (both in terms of its convergence as well as the quality of results) can form a topic for further interesting investigations.

References

Bortolan, G. 1998. An architecture of fuzzy neural networks for linguistic processing. *Fuzzy Sets and Systems*, 100, 197–215.

Buckley, J. and Y. Hayashi. 1994. Fuzzy neural networks: A survey. *Fuzzy Sets and Systems*, 66, 1–14.

Czogała, E. and W. Pedrycz. 1983. On the concept of fuzzy probabilistic controllers. *Fuzzy Sets and Systems*, 10, 109–121.

Hadjili, M.L. and V. Wertz. 2002. Takagi–Sugeno fuzzy modeling incorporating input variables selection. *IEEE Transactions on Fuzzy Systems*, 10, 728–742.

Hirota, K. and W. Pedrycz. 1983. Analysis and synthesis of fuzzy systems by the use of probabilistic sets. *Fuzzy Sets and Systems*, 10, 1–13.

Ishibuchi, H. 1996. Development of fuzzy neural networks. In: *Fuzzy Modelling, Paradigms and Practice*, W. Pedrycz (ed.). Dordrecht: Kluwer Academic Publishers, 185–202.

Ishibuchi, H., K. Kwon, and H. Tanaka. 1995. A learning algorithm of fuzzy neural networks with triangular fuzzy weights. *Fuzzy Sets and Systems*, 71, 277–293.

Ishibuchi, H. and H. Tanaka. 1993. An architecture of neural networks with interval weights and its application to fuzzy regression analysis. *Fuzzy Sets and Systems*, 57, 27–39.

Jafarzadeh, S., M. Fadali, and A. Sonbol. 2011. Stability analysis and control of discrete type-1 and type-2 TSK fuzzy systems: Part I. Stability analysis. *IEEE Transactions on Fuzzy Systems*, 19, 989–1000.

Juang, C. and C. Hsieh. 2012. A fuzzy system constructed by rule generation and iterative linear SVR for antecedent and consequent parameter optimization. *IEEE Transactions on Fuzzy Systems*, 20, 372–384.

Juang, C.F. and C.T. Lin. 1999. A recurrent self-organizing neural fuzzy inference network. *IEEE Transactions on Neural Networks*, 10, 828–845.

Kim, M., C. Kim, and J. Lee. 2006. Evolving compact and interpretable Takagi–Sugeno fuzzy models with a new encoding scheme. *IEEE Transactions on Systems, Man, and Cybernetics, Part B*, 36, 1006–1023.

Kulkolj, D. and E. Levi. 2004. Identification of complex systems based on neural and Takagi–Sugeno fuzzy model. *IEEE Transactions on Systems, Man, and Cybernetics, Part B: Cybernetics*, Vol. 34, 272–282.

Li, C., J. Zhou, B. Fu, P. Kou, and J. Xiao. 2012. T–S fuzzy model identification with a gravitational search-based hyperplane clustering algorithm. *IEEE Transactions on Fuzzy Systems*, Vol. 20, 305–317.

Liu, P. and H. Li. 2004. Efficient learning algorithms for three-layer regular feedforward fuzzy neural networks. *IEEE Transactions on Neural Networks*, 15, 545–558.

Lughofer, E.D. 2008. FLEXFIS: A robust incremental learning approach for evolving Takagi–Sugeno fuzzy models. *IEEE Transactions on Fuzzy Systems*, Vol. 16, 1393–1410.

Moore, R. 1966. *Interval Analysis*. Englewood Cliffs, NJ: Prentice Hall.

Pal, N. and S. Saha. 2008. Simultaneous structure identification and fuzzy rule generation for Takagi–Sugeno models. *IEEE Transactions on Systems, Man, and Cybernetics, Part B*, Vol. 38, 1626–1638.

Park, H., W. Pedrycz, and S.K. Oh. 2009. Granular neural networks and their development through context-based clustering and adjustable dimensionality of receptive fields. *IEEE Transactions on Neural Networks*, 20, 1604–1616.

Pedrycz, W. 1990. Relevancy of fuzzy models. *Information Sciences*, 52, 285–302.

Pedrycz, W. 1998. Shadowed sets: Representing and processing fuzzy sets. *IEEE Trans. on Systems, Man, and Cybernetics, Part B*, 28, 103–109.

Pedrycz, W. and A. Bargiela. Forthcoming 2012. An optimization of allocation of information granularity in the interpretation of data structures: Toward granular fuzzy clustering. *IEEE Transactions on Systems, Man, and Cybernetics, Part B*.

Pedrycz, W., B. Russo, and G. Succi. Forthcoming 2012. Knowledge transfer in system modeling and its realization through an optimal allocation of information granularity. *Applied Soft Computing*.

Pedrycz, W. and A.V. Vasilakos. 1999. Linguistic models and linguistic modelling. *IEEE Trans. on Systems, Man, and Cybernetics*, 29, 745–757.

Pedrycz, W. and G. Vukovich. 2001. Granular neural networks. *Neurocomputing*, 36, 205–224.

Wedge, D., D. Ingram, D. McLean, C. Mingham, and Z. Bandar. 2006. On global–local artificial neural networks for function approximation. *IEEE Transactions on Neural Networks*, 17, 942–952.

Weerdt, E. de, Q.P. Chu, and J.A. Mulder. 2009. Neural network output optimization using interval analysis. *IEEE Transactions on Neural Networks*, 20, 638–653.

Zhang, Y., M.D. Fraser, R.A. Gagliano, and A. Kandel. 2000. Granular neural networks for numerical-linguistic data fusion and knowledge discovery. *IEEE Transactions on Neural Networks*, 11, 658–667.

Zhang, Y., B. Jin, and Y. Tang. 2008. Granular neural networks with evolutionary interval learning. *IEEE Transactions on Fuzzy Systems*, 16, 309–319.

11

Collaborative and Linguistic Models of Decision Making

In group decison making, one strives to reconcile differences of opinions (judgments) expressed by individual members of the group. Fuzzy decision-making mechanisms bring a great deal of flexibility. By admitting membership degrees, we are supplied with a substantial flexibility to exploit different aggregation mechanisms and navigate a process of interaction among decison makers to achieve an increasing level of consistency within the group (consensus building). While the studies reported so far exploit more or less sophisticated ways of adjusting/transforming initial judgments (preferences) of individuals, in this paper we bring forward a concept of information granularity into the overall process of decison making and consensus building. Here information granularity is viewed as an essential asset, which endows a decision maker with a tangible level of flexibility using some initial preferences conveyed by each individual which can be adjusted with the intent of reaching a higher level of consensus. We are concerned with an extension of the well-known analytic hierarchy process (AHP) to the group decision-making scenario. More specifically, in this model the admitted level of granularity gives rise to a *granular* matrix of pairwise comparisons. We also study the problem of moving from linguistic descriptors (which are expressed in a natural language) to their operational mode where the descriptors are associated with some information granules and become essential to any further processing.

11.1 Analytic Hierarchy Process (AHP) Method and Its Granular Generalization

We have been witnessing a highly diversified plethora of approaches to multiperson or group decision-making processes realized in the setting of fuzzy sets. From the very inception of fuzzy sets, these information granules gave rise to numerous concepts, methodologies, algorithms, and applications in human endeavors of decision making. The studies by Zadeh (1975) (and see also Aliev et al., 2004) offer a profound testimony to the relevance of fuzzy sets in diverse tasks of decison making, which are inherently associated with the concept of information granularity. One may refer to Aliev, Fazlollahi, and Aliev (2004) presenting various developments in the models of fuzzy decison making to business and economics. Group decision-making mechanisms exploiting effective ways of forming consensus are on the agenda of fuzzy decision making. Some of these studies build upon the well-known established techniques being formed for a single decision maker. One of the approaches is the analytic hierarchy process (AHP) introduced by Saaty (1987, 1983). Since its introduction there have been a number of studies dealing with its refinements, analysis, generalizations, and applications (Saaty and Ozdemir, 2003; Saaty and Hu, 1998; Wang and Chen, 2008). As one of the generalizations, the AHP method has been augmented to the environment of group decision making, cf. Tung and Tang (1998), McCauley-Bell and Badiru (1996), Mustafa and Al-Bahar (1991). A quick browsing of the existing studies in this setting helps us establish a general taxonomy of approaches with two dominant directions:

Aggregation of individual reciprocal matrices (individual judgments). Here the focus is on forming an aggregate of the reciprocal matrices. For instance, a weighted geometric mean method is used as a vehicle to realize this aggregation.

Aggregation of individual vectors of preferences produced by the AHP method. The preferences obtained on the basis of the individual reciprocal matrices are then subject to a certain aggregation.

There have been many studies elaborating on the advantages and associated limitations of the aggregation methods originating from these two categories (Cakir, 2008; Chu and Liu, 2002; Dong et al., 2010; Frei and Harker, 1999).

An interesting and practically legitimate direction of promising investigations is devoted to group decision making involving linguistic or generally speaking, nonnumeric representations of assessments of individual alternatives (Herrera et al., 1997, 2005, 2007). In all these situations, it is assumed that such types of representation of objectives, criteria, and their relevance are reflective of the very nature of knowledge about the decision problem. It is also assumed that such nonnumeric quantifications are easily provided by the decision maker.

One can convincingly note that reaching consensus requires flexibility and willingness on the part of each member of the group to adjust his/her initial position. In the AHP model, these changes can be articulated through the modifications of the entries of the individual reciprocal matrices. Intuitively, if we allow any decison maker to treat the entries of the reciprocal matrices not as single numeric values (which are rigid and difficult to adjust) but rather as information granules: this will bring a badly needed factor of flexibility. In this manner the decision maker may admit all numeric realizations of the reciprocal matrix, which are compatible with the more general granular abstraction such as the assumed granular matrix of pairwise comparison. We introduce a concept of granular reciprocal matrices and emphasize a role of information granularity being regarded here as an important conceptual and computational resource which can be exploited as a means to elevate the level of consensus to be reached. In a nutshell, the level of granularity could be treated as synonymous with the level of flexibility injected into the modeling environment. This terminology underlines an important, unique role being played by information granules. By admitting a certain level of information granularity, we are provided with a unique possibility to navigate within the space of priorities assigned to the individual alternatives. In a very descriptive and informal way, the essence of the information granularity is to support reaching consensus through bringing some flexibility and exploiting it to the fullest possible extent.

11.2 Analytic Hierarchy Process Model—The Concept

The analytic hierarchy process (AHP) developed in a series of studies by Saaty (1983, 1987) and further researched and generalized by others (Jeonghwan et al., 2010; Kangmao, Wang, and Chun, 2005; Korpela and Tuominen, 1996) is aimed at forming a vector of preferences given a finite set of alternatives. Let us consider a collection of "n" alternatives $x_1, x_2, ..., x_n$. These preferences are formed on the basis of a reciprocal matrix R, R = $[r_{ij}]$, i, j = 1, 2, ..., n whose entries are a result of pairwise comparisons of alternatives provided by a decision maker. The starting point of the estimation process of the fuzzy set of preferences are entries of the reciprocal matrix which are obtained through collecting results of pairwise evaluations offered by an expert, designer, or user (depending on the character of the task at hand). Prior to making any assessment, the expert is provided with a finite scale with integer values in-between 1 and 7. Some other alternatives of the scales such as those involving 5 or 9 levels could be sought as well. If x_i is strongly preferred over x_j when being considered in the context of the fuzzy set whose membership function we would like to estimate, then this judgment is expressed by assigning high values of the available scale, that is, 6 or 7. If we still sense that x_i is preferred over a_j, yet the strength of this preference is lower in comparison with the previous case, then this is quantified using some intermediate values of the scale, that is, 3 or 4. If no difference is considered, the values close to 1 are the preferred choice, that is, 2 or 1. The value of 1 indicates that x_i and x_j are equally preferred. The general quantification of preferences positioned on the scale of 1 to 9 can be described as follows

Equal importance: 1

Moderate importance of one element over another: 3

Strong importance: 5

Demonstrated importance: 7

Extreme importance: 9

There are also some intermediate values, which could be used to further quantify the relative dominance. On the other hand, if x_j is preferred over x_i, the corresponding entry assumes values below one. Given the reciprocal nature of the assessment, once the preference of a_i over a_j has been quantified, the inverse of this number is inserted into the entry of the matrix that is located at the (j, i)-th coordinate. Next, the maximal eigenvalue λ_{max} is computed along with the corresponding eigenvector. The normalized version of the eigenvector is then regarded as the membership function of the fuzzy set: we arrive at the set of alternatives studied in the problem when realizing all pairwise assessments of the elements of its universe of discourse. The effort to complete pairwise evaluations is far more manageable in comparison to any experimental overhead we require when assigning membership grades to all elements (alternatives) at the same time.

Another advantage comes with an ability to assess the consistency of the pairwise comparisons. In an ideal case, the largest eigenvalue is equal to the number of alternatives. When we encounter some level of inconsistency (which is a reflection of some violation of the transitivity property), then the eigenvalue gets higher. A level of inconsistency of evaluations (the reciprocal matrix) is expressed in terms of the following inconsistency index

$$v = \frac{\lambda_{max} - n}{n - 1} \tag{11.1}$$

The higher the value of this index is, the more significant level of inconsistency is associated with the preferences collected in the reciprocal matrix. The values of v around 0.2 or 0.1 are often regarded as acceptable threshold levels. If v exceeds the thresholds, the results of pairwise comparison are deemed inconsistent, which requires further refinements of the assessments of some pairwise comparisons or a repetition of the entire experiment. It is worth noting that any value of threshold needs to be cast in the context of a certain application (refer also to an interesting discussion on this issue covered in Apostolou and Hassell, 2002). The threshold of the consistency ratio (CR) expressed as the ratio of the consistency index (CI) and the random consistency index (RI), CR = CI/RI is also established (typically assuming the value of 0.1) to assess the quality of the results of pairwise comparison.

11.3 Granular Reciprocal Matrices

Consider a group decision-making scenario in which there are "c" decison makers and each of them comes with his/her own preferences (preference vectors), e[1], e[2], ..., e[c] obtained by running the AHP for the corresponding reciprocal matrix R[1], R[2], ..., R[c]. Furthermore the quality of preference vectors is quantified by the associated inconsistency index v[i].

An aggregate preference vector can be formed in many different ways by taking into account various methods of aggregation of R[i]s or **e**[i]s and accounting for the corresponding levels of consistency. This issue was elaborated on in the introductory section. It is worth noting that regardless of the aggregation technique being used, the resulting vector of preferences is somewhat the result of a postmortem processing in which the individual decision makers are rather passive to changes/adjustments of their initial positions expressed in the form of the corresponding vectors of preference. The question of aggregation is interesting per se as so many distinct scenarios can be encountered. For instance, there is a group of decision makers who share a similar position about preferences and each of them is not very consistent. A few of the members of the group who are in a clearly visible minority have very different preferences, however, their reciprocal matrices are highly consistent. A number of interesting problems arise. How could this become reflected in the aggregation mechanism? What if there are several subgroups of decision makers and each group is quite consistent in articulating their preferences but these preferences vary substantially from one subgroup to another?

Building consensus calls for some flexibility exhibited by all decision makers who in the name of cooperative pursuits give up their initial positions and show a certain level of elasticity. They can demonstrate this by changing the original entries of the individual reciprocal matrices. The very much needed elasticity/flexibility is brought into the AHP structures by allowing the reciprocal matrices to be *granular* rather than numeric. By granular reciprocal matrices we mean matrices whose entries are not plain numbers but information granules, that is, intervals, fuzzy sets, rough sets, probability density functions, and others. In a nutshell, information granularity present here serves as an important modeling asset, which brings forward an ability of the decision maker to exercise some flexibility to be used in modifying his/her own position when becoming aware of the preferences of the other members of the group. In essence, the reciprocal matrix is elevated (abstracted) to its granular format. To emphasize this, we use the notation $G(R)$ to underline the fact that we are concerned with granular reciprocal matrices where $G(.)$ stands for a specific granular formalism being used here. Being more specific, we can talk about interval-valued reciprocal matrices.

Evidently, the higher the level of granularity that is provided to the individual decison maker, the higher the likelihood of arriving at preferences accepted by all members of the group. Here we appeal to the intuitive concept of granularity by trying to present a qualitative nature of the process in which the asset of granularity is involved. This idea can be formalized depending on the form of the information granules being the entries of the reciprocal matrices. For instance, if the granularity of information is articulated through intervals, the length of such intervals—entries of the reciprocal matrix can be sought as a level of granularity. In the case of fuzzy sets, the level of granularity can be gauged by σ-counts of fuzzy sets (fuzzy numbers) that are the entries of the granular (fuzzy) reciprocal matrix. In our further considerations, we confine ourselves to interval-valued reciprocal matrices with the level of granularity equal to α. The flexibility offered by the level of granularity can be effectively used to optimize a certain objective function capturing the essence of the reconciliation of the individual preferences.

The idea of having entries of the reciprocal matrix R treated as granular entities was originally introduced in Van Laarhoven and Pedrycz (1983). More specifically, in this study a generalized version of the AHP was discussed where the reciprocal matrices were composed of triangular fuzzy numbers. The underlying motivation, different from the one presented here, was to reflect uncertainty present in judgments of decision makers when forming pairwise estimates. The one advocated here exhibits some resemblance in the sense we are again concerned with the idea of granularity of information. The crucial difference lies in the treatment of information granularity—here we treat it as a conceptual vehicle to facilitate admissible changes to the results of pairwise comparison.

11.3.1 The Objective Function

It is advantageous to look at the way in which the overall process of forming consensus can be structured. Figure 11.1 shows a characterization of the preferences at the level of the individual decision maker and a way

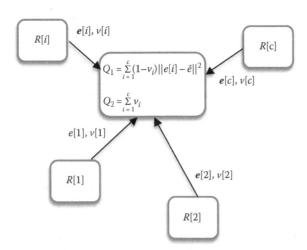

FIGURE 11.1
Characterization of preferences at the level of individual decision makers and a way of forming the objective function (see a detailed description in the text).

in which an entire process of consensus building becomes realized at the group level.

In the interval-valued granular model of reciprocal matrices, we consider that the individual decision maker feels equally comfortable to choose any reciprocal matrix whose values are located within the bounds established by the predetermined level of granularity. If a certain level of granularity (α) is allowed in each reciprocal matrix, it can be used in two ways to reach consensus.

First, we try to bring all preferences close to each other and this goal is realized by adjusting the reciprocal matrices within the bounds offered by the admissible level of granularity provided to each decision maker.

$$Q_1 = \sum_{i=1}^{c} (1 - v_i) \, ||e[i] - \hat{e}||^2 \tag{11.2}$$

where $\hat{e}$ stands for the vector of preferences which minimizes the weighted sum of differences $||.||$ between $e[i]$ and $\hat{e}$. The detailed calculations depend on the form of the distance function used here. In particular, if we select the Euclidean distance, then the optimal vector of preferences reads as follows:

$$\hat{e} = \frac{\sum_{i=1}^{c} (1 - v_i) e_i}{\sum_{i=1}^{c} (1 - v_i)} \tag{11.3}$$

Second, we try to increase the consistency of the reciprocal matrices and this improvement is realized at the level of individual decision maker. The following performance index quantifies this effect,

$$Q_2 = \sum_{i=1}^{c} v_i \tag{11.4}$$

These are the two objectives to be minimized. If we consider the scalar version of the optimization problem, it arises in the following form:

$$Q = AQ_1 + Q_2 \tag{11.5}$$

where $A \geq 0$. The overall optimization problem reads now as follows:

$$\text{Min}_{R[1], R[2], \ldots, R[c] \in P(R)} Q \tag{11.6}$$

The minimization above is carried out for all reciprocal matrices admissible because of the introduced level of information granularity α. This fact is underlined by including a granular form of the reciprocal matrices allowed in the problem, that is, R[1] R[2], ..., R[c] are elements of the family of interval-valued reciprocal matrices, namely, $P(R)$.

The selection of numeric values of the weight factor A and the level of granularity α require some clarification. The role of A is quite straightforward: by choosing its value we set up a tradeoff between the consistency obtained at the local (individual decision maker) and global group level. The higher the value of A, the more attention is being paid to the consistency aimed at the group level. In the limit where $A = 0$, we are concerned with the consistency achieved at the local level only. The admissible level of granularity brings flexibility, which has to be effectively utilized—the higher the values of α, the higher the potential to reach a significant level of consistency. There is a potential of producing some quite inconsistent reciprocal matrices at the local level.

The convincing generalization of the optimization problem discussed above is to optimize allocation of granularity to the individual decision makers. While in the above scenario we considered that the same level of granularity α is assigned across all R[i]s, these levels can be optimized as well so that each decision maker might have an individual value of a_i that becomes available to his disposal. We require that the overall balance of granularity is retained meaning that the following requirement

$$c\alpha = \sum_{i=1}^{c} \alpha_i \qquad (11.7)$$

is satisfied. The values of $\alpha_1, \alpha_2, \ldots, \alpha_c$ become a part of the overall optimization problem, that is, Equation (11.6) is expanded and reads now as follows:

$$\text{Min}_{R[1], R[2], \ldots, R[c] \in P(R)} Q$$

subject to the granularity constraint Equation (11.7). Figure 11.2 visualizes the essential role of information granularity in endowing the reciprocal matrices with flexibility.

The optimization of the reciprocal matrices coming from the space of granular reciprocal matrices (more precisely, interval-valued reciprocal matrices) is realized by means of the particle swarm optimization (PSO), which occurs as a viable optimization alternative for this problem. The PSO is well documented in the existing literature with numerous modifications and augmentations. The reader may refer to the generic flow of computing in which velocities and positions of the particles are updated. What is important in this setting is a formation of the particle whose entries are transformed to the values of the reciprocal matrix. From the perspective of the PSO itself, all entries of the particle are located in the unit interval. As the matrix exhibits the reciprocity property, only one of the entries located symmetrically with respect to the main diagonal has to optimized by the PSO and the other one is determined on the basis of the reciprocity property.

Starting with the initial reciprocal matrix provided by the expert and assuming a given level of granularity α (located in the [0,1] interval), let us

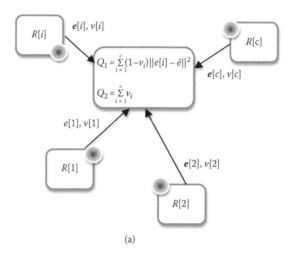

(a)

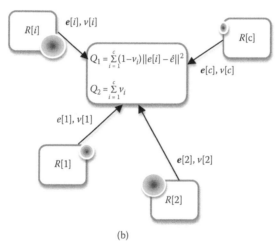

(b)

FIGURE 11.2
Endowing the reciprocal matrices with flexibility represented in the form of information granularity. (a) Equal allocation of granularity—level of granularity set to α. (b) Optimal allocation of granularity with different level α_i assigned to individual reciprocal matrices under the constraint of overall level of granularity expressed by Equation (11.7). Shadowed shapes positioned next to the reciprocal matrices underline the granular nature of these constructs.

consider an entry for which r_{ij} is lower than 1. The interval of admissible values of this entry of $P(R)$ implied by the level of granularity is equal to

$$[a, b] = [\max(1/9, r_{ij} - a(8/9)) , \min(1, r_{ij} + a(8/9))] \tag{11.8}$$

Let us assume that a certain entry of interest of the particle is "x." It is transformed linearly according to the expression, $z = a + (b - a)x$. We compute its inverse, $1/z$, and round it off to the nearest integer in the set $\{1, ..., 9\}$. Suppose the result is the integer value z_0. To retain the reciprocality property, we calculate the inverse of z_0 and replace the result of the linear transformation (z) from which we have started with. For example, consider that r_{ij} is equal to $1/5$, the admissible level of granularity $\alpha = 0.05$ and the corresponding entry of the particle is $x = 0.3$. Then the corresponding interval of the granular pairwise comparison matrix computed as given by Equation (11.8) becomes equal to $[a, b] = [0.155, 0.244]$. Subsequently $z = 0.181$. This gives rise to the integer value of the (j, i)-th entry of R positioned symmetrically equal to 6 and finally the modified value of r_{ij} becomes equal to $1/6$.

The overall particle is composed of the individual segments where each of them is concerned with the optimization of the parameters of the reciprocal matrices. If the allocation of granularity is also optimized then the values of a_is are also optimized. These optimized values are then used in the decoding of the content of the corresponding segments of the particle by calibrating the intervals of admissible granularity. More specifically, for the optimized values of R[i] we use the interval $[\max(1/9, r_{lj} - \alpha_i(8/9))$, $\min(1, r_{lj} + \alpha_i(8/9))]$ where r_{lj} is the (l, j)-th element of the reciprocal matrix.

Example

We consider five reciprocal matrices coming from five decison makers. The entries of these matrices are reflective of the pairwise comparisons of five alternatives (options)

$$
R_1 = \begin{bmatrix}
1 & 1/4 & 1/3 & 1/7 & 1/3 \\
4 & 1 & 1/6 & 1/5 & 1/6 \\
3 & 6 & 1 & 1/2 & 1/3 \\
7 & 5 & 2 & 1 & 1/2 \\
3 & 6 & 3 & 2 & 1
\end{bmatrix}
\quad
R_2 = \begin{bmatrix}
1 & 1/2 & 1/4 & 1/3 & 1/2 \\
2 & 1 & 1/5 & 1/2 & 2 \\
4 & 5 & 1 & 1 & 1/2 \\
3 & 2 & 1 & 1 & 3 \\
2 & 1/2 & 2 & 1/3 & 1
\end{bmatrix}
$$

$$
R_3 = \begin{bmatrix}
1 & 1/2 & 1/2 & 1/8 & 1/8 \\
2 & 1 & 1/4 & 1/3 & 1/9 \\
2 & 4 & 1 & 1/5 & 1/5 \\
8 & 3 & 5 & 1 & 1/4 \\
8 & 9 & 5 & 4 & 1
\end{bmatrix}
\quad
R_4 = \begin{bmatrix}
1 & 1/2 & 1/3 & 1/7 & 1/2 \\
2 & 1 & 1/5 & 1/2 & 1/3 \\
3 & 5 & 1 & 1/4 & 1/4 \\
7 & 2 & 4 & 1 & 1 \\
2 & 3 & 4 & 1 & 1
\end{bmatrix}
$$

$$
R_5 = \begin{bmatrix}
1 & 1 & 3 & 5 & 5 \\
1 & 1 & 1/3 & 4 & 8 \\
1/3 & 3 & 1 & 3 & 6 \\
1/5 & 1/4 & 1/3 & 1 & 6 \\
1/5 & 1/8 & 1/6 & 1/6 & 1
\end{bmatrix}
$$

The corresponding maximal eigenvalues of the reciprocal matrices are equal to: $\lambda_1 = 5.68$, $\lambda_2 = 5.79$, $\lambda_3 = 5.46$, $\lambda_4 = 5.73$, $\lambda_5 = 5.70$, which give rise to the following values of the inconsistency index $v_1 = 0.171$, $v_2 = 0.198$, $v_3 = 0.114$, $v_4 = 0.183$, $v_5 = 0.175$. It is noticeable that all the matrices exhibit a similar level of consistency with an exception of the reciprocal matrix R_3, whose consistency level is higher than for the rest of the matrices. The corresponding eigenvectors are given below

$$\mathbf{e}_1 = \begin{bmatrix} 0.141 & 0.206 & 0.506 & 0.753 & 1.000 \end{bmatrix}^\mathrm{T}$$

$$\mathbf{e}_2 = \begin{bmatrix} 0.239 & 0.509 & 1.000 & 0.970 & 0.638 \end{bmatrix}^\mathrm{T}$$

$$\mathbf{e}_3 = \begin{bmatrix} 0.075 & 0.106 & 0.195 & 0.481 & 1.000 \end{bmatrix}^\mathrm{T}$$

$$\mathbf{e}_4 = \begin{bmatrix} 0.187 & 0.268 & 0.501 & 1.000 & 0.884 \end{bmatrix}^\mathrm{T}$$

$$\mathbf{e}_5 = \begin{bmatrix} 1.000 & 0.647 & 0.768 & 0.254 & 0.096 \end{bmatrix}^\mathrm{T}$$

There is a diversity of preferences discovered here: the fifth alternative is picked up twice and the remaining alternatives are identified once. In case no granularity is admitted, the aggregated fuzzy set of preferences computed by making use of Equation (11.3) comes with a vector of membership values $\hat{\mathbf{e}} = [0.3255 \quad 0.3427 \quad 0.5861 \quad 0.6867 \quad 0.728]^\mathrm{T}$.

Example

Here we consider a larger group of decision makers who have quantified the pairwise preferences of five alternatives in the following matrices:

$$R_1 = \begin{bmatrix} 1 & 1/7 & 1/5 & 1/4 & 1 \\ 7 & 1 & 4 & 2 & 9 \\ 5 & 1/4 & 1 & 1/2 & 3 \\ 4 & 1/2 & 2 & 1 & 7 \\ 1 & 1/9 & 1/3 & 1/7 & 1 \end{bmatrix} \quad R_2 = \begin{bmatrix} 1 & 9 & 7 & 6 & 1 \\ 1/9 & 1 & 1/3 & 1/2 & 1/7 \\ 1/7 & 3 & 1 & 2 & 1/5 \\ 1/6 & 2 & 1/2 & 1 & 1/6 \\ 1 & 7 & 5 & 6 & 1 \end{bmatrix}$$

$$R_3 = \begin{bmatrix} 1 & 1/6 & 1/5 & 1 & 1/7 \\ 6 & 1 & 2 & 7 & 1/2 \\ 5 & 1/2 & 1 & 9 & 1/3 \\ 1 & 1/7 & 1/9 & 1 & 1/9 \\ 7 & 2 & 3 & 9 & 1 \end{bmatrix} \quad R_4 = \begin{bmatrix} 1 & 5 & 8 & 1/3 & 1/2 \\ 1/5 & 1 & 1 & 1/6 & 1/7 \\ 1/8 & 1 & 1 & 1/9 & 1/8 \\ 3 & 6 & 9 & 1 & 2 \\ 2 & 7 & 8 & 1/2 & 1 \end{bmatrix}$$

$$R_5 = \begin{bmatrix} 1 & 1/2 & 1/5 & 8 & 1 \\ 2 & 1 & 1/4 & 5 & 1/3 \\ 5 & 4 & 1 & 6 & 4 \\ 1/8 & 1/5 & 1/6 & 1 & 1/8 \\ 1 & 3 & 1/4 & 8 & 1 \end{bmatrix} \quad R_6 = \begin{bmatrix} 1 & 8 & 1/7 & 1/6 & 1 \\ 1/8 & 1 & 1/8 & 1/5 & 1/2 \\ 7 & 8 & 1 & 3 & 9 \\ 6 & 5 & 1/3 & 1 & 5 \\ 1 & 2 & 1/9 & 1/5 & 1 \end{bmatrix}$$

$$R_7 = \begin{bmatrix} 1 & 1/7 & 1/6 & 1 & 2 \\ 7 & 1 & 1/3 & 5 & 3 \\ 6 & 3 & 1 & 8 & 9 \\ 1 & 1/5 & 1/8 & 1 & 6 \\ 1/2 & 1/3 & 1/9 & 1/6 & 1 \end{bmatrix} \quad R_8 = \begin{bmatrix} 1 & 6 & 1/7 & 1/4 & 1 \\ 1/6 & 1 & 1/9 & 1/5 & 2 \\ 7 & 9 & 1 & 3 & 7 \\ 4 & 5 & 1/3 & 1 & 5 \\ 1 & 1/2 & 1/7 & 1/5 & 1 \end{bmatrix}$$

For these matrices we obtain the following set of eigenvalues and the corresponding values of the inconsistency index:

$\lambda_1 = 5.1180$, $\lambda_2 = 5.1177$, $\lambda_3 = 5.1605$, $\lambda_4 = 5.1599$, $\lambda_5 = 5.5979$, $\lambda_6 = 5.5992$, $\lambda_7 = 5.5988$, $\lambda_8 = 5.6007$; $v_1 = 0.0295$, $v_2 = 0.0294$, $v_3 = 0.0401$, $v_4 = 0.04$, $v_5 = 0.1495$, $v_6 = 0.1498$, $v_7 = 0.1497$, $v_8 = 0.1502$ along with the eigenvalues describing levels of preference assigned to the alternatives

$$e_1 = \begin{bmatrix} 0.109 & 1.000 & 0.334 & 0.559 & 0.100 \end{bmatrix}^T \quad e_2 = \begin{bmatrix} 1.000 & 0.093 & 0.217 & 0.148 & 0.879 \end{bmatrix}^T$$

$$e_3 = \begin{bmatrix} 0.103 & 0.636 & 0.466 & 0.086 & 1.000 \end{bmatrix}^T \quad e_4 = \begin{bmatrix} 0.457 & 0.106 & 0.087 & 1.000 & 0.686 \end{bmatrix}^T$$

$$e_5 = \begin{bmatrix} 0.282 & 0.280 & 1.000 & 0.067 & 0.416 \end{bmatrix}^T \quad e_6 = \begin{bmatrix} 0.196 & 0.069 & 1.000 & 0.533 & 0.120 \end{bmatrix}^T$$

$$e_7 = \begin{bmatrix} 0.129 & 0.527 & 1.000 & 0.187 & 0.083 \end{bmatrix}^T \quad e_8 = \begin{bmatrix} 0.212 & 0.100 & 1.000 & 0.484 & 0.109 \end{bmatrix}^T$$

After running the PSO algorithm, the results obtained are reported in Figure 11.3. The cumulative level of inconsistency (expressed by Q_2) drops very quickly with the increasing values of granularity, Figure 11.3a, and this effect is noticeable for different values of A. The values of the performance index Q_1 are reduced, however, the trend is not so visible or not present as in the case of Q_2.

11.4 A Quantification (Granulation) of Linguistic Terms as Their Operational Realization

The linguistic terms used in a pairwise comparison of alternatives are expressed linguistically by admitting qualitative terms. They can be organized in a linear fashion, as there is some apparent linear order among them. The terms themselves are not operational meaning that no further detailed processing can be realized, which involves a quantification of the linguistic terms. Schematically, we can portray the process of arriving at the operational representation of linguistic terms as illustrated in Figure 11.4. In this figure, capital letters denote the corresponding linguistic terms: L (*low*), M (*medium*), and H (*high*).

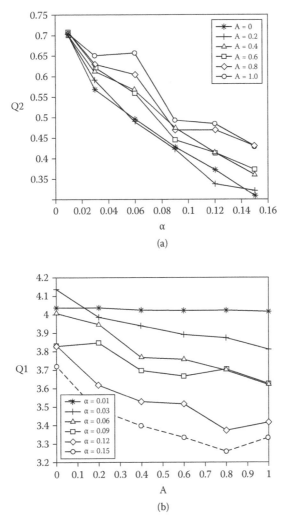

FIGURE 11.3
Q_2 versus α and Q_1 versus A.

The two important features of such granulation mechanisms are worth noting here: (a) the mapping is by no means linear, that is, a localization of the associated information granules on the scale is not uniform, (b) the semantics of the terms allocated in the process of granulation is retained. As usual, various information granulation formalisms can be contemplated including sets (intervals), rough sets, fuzzy sets, and shadowed sets.

The question of how to arrive at the operational version of the information granules can be reformulated as a certain optimization problem and solved with the use of the PSO. We elaborate on the pertinent fitness function, its realization, and the optimization details.

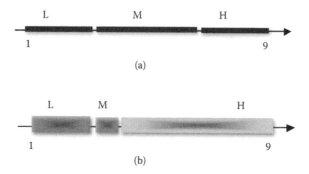

FIGURE 11.4
Toward the operational realization of linguistic terms. (a) Realization with the aid of intervals. (b) Fuzzy set-based implementation.

11.4.1 Evaluation of the Mapping from Linguistic Terms to Information Granules

The objective of the fitness function is to provide a quantification of the information granules on which information granules are to be mapped. Considering the nature of the AHP model, the quality of the solution (preference vector) is expressed in terms of the inconsistency index. For the given vector of cutoff points, their quality associates with the corresponding value of the inconsistency index. The minimization of the values of the index by adjusting the positions of the cutoff points in the 1 ... 9 scale is realized by the PSO. When it comes to the formation of the fitness function, its determination has to take into account the fact that interval-valued entries of the reciprocal matrix have to return numeric values of the fitness function. This is realized as follows. As we encounter information granules in the form of intervals, we consider a series of their realizations. Let us consider that an individual generated by the PSO has produced a collection of cutoff points specified by the individual in the swarm on which PSO operate, which is located on the [1, 9] scale. For instance, if there are three linguistic terms in the reciprocal matrix, that is, L (*Low*), M (*Medium*), and H (*High*), and the corresponding cutoff points are a_1 and a_2, respectively (see Figure 11.4), we arrive at the mapping (representation) of the terms as follows L : [1, a_1] M: [a_1, a_2] H: [a_2, 9]. If we consider "m" linguistic values, this results in "m-1" cutoff points. Being arranged in a single vector, they constitute an individual in the swarm of the PSO.

A finite series of realizations of the information granules being the entries of the granular reciprocal matrix is formed by randomly generating entries coming from the above intervals, plugging them into the reciprocal matrix, and computing the largest eigenvalue and the corresponding value of the inconsistency index. The average of the values of the inconsistency index is the fitness function associated with the particle formed by the cutoff points a_1 and a_2. This way the formation of the fitness function is in line with the standard practices encountered in Monte Carlo simulations.

Example

We consider a 5 x 5 reciprocal matrix with the three linguistic entries

$$R = \begin{bmatrix} 1 & H & L & 1/L & 1/L \\ 1/H & 1 & 1/L & 1/M & 1/M \\ 1/L & L & 1 & 1/M & 1/M \\ L & M & M & 1 & 1/L \\ L & M & M & L & 1 \end{bmatrix}$$

The granular matrix R is sampled 500 times (the numbers drawn from the uniform distribution defined over the corresponding subintervals of the [1,9] scale). Recall that the fitness function is the average of the inconsistency index computed over each collection of 500 reciprocal matrices. The process of learning is realized by the PSO. The progression of the optimization is quantified in terms of the fitness function obtained in successive generations (see Figure 11.5).

To put the obtained optimization results in a certain perspective, we report the performance obtained when considering a uniform distribution of the cutoff points over the scale, which are equal to 3.67 and 6.34,

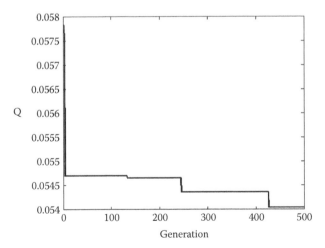

FIGURE 11.5
The values of the fitness versus successive generations of the PSO.

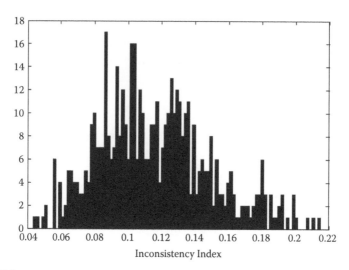

FIGURE 11.6
The distribution of the values of the inconsistency index—a uniform distribution of the cutoff points.

respectively. The average inconsistency index assumes the value of 0.1159 with a standard deviation of 0.0332. The histogram of the inconsistency rates provides a more comprehensive and convincing view of the results: there is a visible presence of a longer tail of the distribution stretched toward higher values of the inconsistency index (Figure 11.6).

The PSO returns the optimal cutoff points of 2.2 and 2.4, which are evidently shifted toward the lower end of the scale. The inconsistency index now takes on lower values and is equal to 0.054 with the standard deviation of 0.0252. The corresponding histogram is shown in Figure 11.7.

For the optimal splits of the scale, a reciprocal matrix with the lowest inconsistency index is given below,

$$
R = \begin{bmatrix}
1 & 2.2 & 1/1.65 & 1/2.35 & 1/2.4 \\
1/2.2 & 1 & 1/2.1 & 1/4.05 & 1/6.15 \\
1.65 & 2.1 & 1 & 1/1.45 & 1/2.45 \\
2.35 & 4.05 & 1.45 & 1 & 1/1.25 \\
2.4 & 6.15 & 2.45 & 1.25 & 1
\end{bmatrix}
$$

with the normalized eigenvector corresponding to the largest eigenvalue of this reciprocal matrix equal to $\mathbf{e} = [0.79 \ 1.00 \ 0.65 \ 0.30 \ 0.00]^T$, which identifies the second alternative as an optimal one.

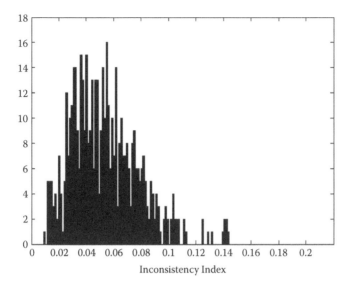

FIGURE 11.7
The distribution of the values of the inconsistency index—a PSO-optimized distribution of the cutoff points.

Example

Here we consider another 5 x 5 reciprocal matrix with five linguistic terms, VL, L, M, H, and VH

$$
R = \begin{bmatrix}
1 & L & M & VL & H \\
1/L & 1 & M & 1/VL & VH \\
1/M & 1/M & 1 & 1/H & L \\
1/VL & VL & H & 1 & H \\
1/H & 1/VH & 1/L & 1/H & 1
\end{bmatrix}
$$

The results of the optimization are shown in Figure 11.8, here the parameters of the PSO were set up as follows: number of particles: 100, number of iterations: 500, $c_1 = c_2 = 2$.

The results corresponding with the uniform distribution of the cutoff points (that is, 2.6, 4.2, 5.8, and 7.4) come with the average inconsistency index of 0.0888 with a standard deviation of 0.0162. In contrast, the PSO produces the cutoff points of 1.1, 1.8, 5.3, and 5.6. The inconsistency index is now lower and equal to 0.0205 with the standard deviation of 0.0102. The corresponding histograms both for the uniform and PSO-optimized cutoff points are shown in Figure 11.9.

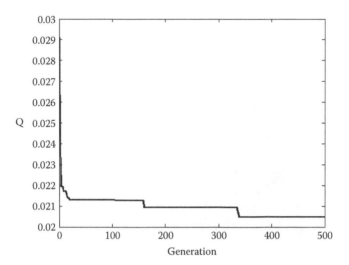

FIGURE 11.8
The values of the fitness function obtained in successive generations of the PSO.

For the optimal split of the scale, a reciprocal matrix with the lowest inconsistency index is given below,

$$
R = \begin{bmatrix}
1 & 1.1 & 4.55 & 1.05 & 5.5 \\
1/1.1 & 1 & 3.85 & 1/1.05 & 6.05 \\
1/4.55 & 1/3.85 & 1 & 1/5.3 & 1.3 \\
1/1.05 & 1.05 & 5.3 & 1 & 5.3 \\
1/5.5 & 1/6.05 & 1/1.3 & 1/5.3 & 1
\end{bmatrix}
$$

with the corresponding eigenvalue equal to $\mathbf{e} = [1.00\ \ 0.92\ \ 0.06\ \ 0.97\ \ 0.00]^{T}$.

11.5 Granular Logic Operators

In what follows, we introduce the concept of a granular representation of numeric membership functions of fuzzy sets, which offers a synthetic and qualitative view of fuzzy sets and their ensuing processing. The notion of consistency of the granular representation is formed, which helps regard the problem as a certain optimization task. More specifically, the consistency is referred to a certain operation φ, which gives rise to the concept of φ-consistency. Likewise introduced is a concept of granular consistency with regard to a collection of several operations. Given the essential role played by

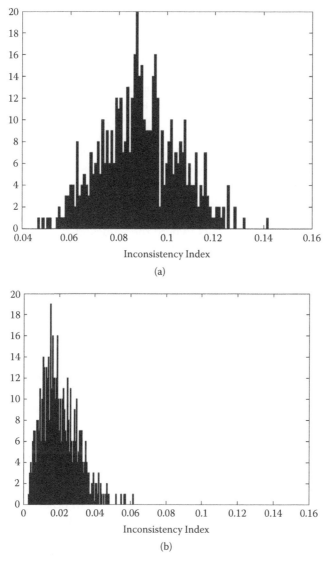

FIGURE 11.9
The distribution of the inconsistency index. (a) Uniform distribution of cutoff points. (b) PSO-optimized distribution of the cutoff points.

logic operators in computing with fuzzy sets, detailed investigations include *and*- and *or*-consistency as well as (*and*- , *or*)-consistency of granular representations of membership functions with the logic operators implemented in the form of various t-norms and t-conorms. The optimization framework supporting the realization of the f-consistent optimization process is provided through particle swarm optimization.

Just recently, there has been a growing interest in taking a more general view of membership grades regarded as nonnumeric quantities; refer to the most recent study (Zadeh, 2011) lucidly advocating this point. Interval fuzzy sets or type-2 fuzzy sets (Mizumoto and Tanaka, 1981) can be viewed as another manifestation of the nonnumeric perspective of membership functions. Some augmentations of the logic operators stress a statistical facet of processing behind processing fuzzy sets, as emphasized through a construct of statistically grounded logic operators (Pedrycz, 2009). Such statistically grounded logic operators incorporate the nature of the data as a part of the logic constructs of operators. In spite of the profound diversity of the existing pursuits, all of these logic operations share a visible commonality: they process numeric membership functions, and, as a result produce detailed numeric results. The study presented in Pedrycz (2009) helps focus on the nonnumeric view of membership functions by representing (approximating) them by several selected α-cuts (where the values of the thresholds themselves are subject to the optimization).

The ultimate objective is to develop a granular representation of membership functions in such a way that the granular membership functions (granular manifestations of numeric membership functions) coincide with (approximate or represent) the original membership function. More formally, the problem is formulated as follows. Consider two fuzzy sets with membership grades A and B assumed for a given element of the universe of discourse $\mathbf{X}$. Given is a certain numeric transformation of A and B, denoted here by ϕ, which produces the membership grade $C = \phi(A, B)$. In particular, ϕ can be a certain logic operator. A granulation of membership functions G is a way of representing the unit interval of membership values as a finite and small collection of information granules such as sets, fuzzy sets, rough sets, shadowed sets, and so forth. These information granules come with a well-defined semantics, that is, *Low, Medium, High, Very High* membership, and so forth. A vocabulary comprising a finite number of information granules coming as a result of granulation $G(A)$ is used as a granular representation of the original numeric membership grade A. In other words, instead of talking about numeric membership values, that is, 0.74, 0.236, and so on, we are concerned about a more abstract view of the membership concept. The underlying task is to construct such information granules (intervals, fuzzy sets, etc.)—components of the vocabulary $\{L_1, L_2, \dots L_c\}$ so that the following relationship

$$G(f(A, B)) \approx \phi\, (G(A), G(B)) \tag{11.9}$$

becomes satisfied to the highest extent. The detailed meaning of the "approximately equal" relationship ($\approx$) shown above will be clarified later on. We show that a solution to Equation (11.9) entails a certain optimization problem. As the transformation (mapping) ϕ appears explicitly in the construction of the information granules—granular descriptors of fuzzy sets, we will be referring to such descriptors as ϕ-consistent. For instance, we can talk about *and, or, complement, match*—consistent granular descriptors of fuzzy sets, which relate to corresponding logic operators (*and, or*) or some matching operations. Likewise, in the case of several operators being considered at the same time (*en block*), that is, ϕ, γ, η ... we arrive at (ϕ, γ, η...)-consistent granular description of fuzzy sets. For instance, in this way we can talk about (*and, or*)-consistent granular description (representation) of fuzzy sets.

11.5.1 Construction of Information Granules of Membership Function Representation: An Optimization Problem

The problem of a granular representation or description of fuzzy sets is concerned with the formation of a family of information granules over the unit interval—a granular range of membership values. In this study, we consider an interval format of information granulation meaning that the information granules come in the form of intervals $[a_k, a_{k+1}] \subset [0,1]$, namely, information granules $L_1, L_2, ..., L_c$ where $L_1 = [0, a_1)$, $L_2 = [a_1, a_2)$... $L_i = [a_{k-1}, a_k)$... $L_c = [a_{c-1}, 1]$. The above intervals form a partition of the unit interval where $0 < a_1 < ... < a_{c-1} < 1$. The interval format of granulation of the unit interval is fully characterized by the vector of parameters of the granular transformation of the unit interval, $\mathbf{a} = [a_1\ a_2\ ..a_{c-1}]^T$.

Before moving on to the general concept and ensuing optimization problems, let us start with an illustrative example highlighting the essence of the construct. When applying the granulation process (granulation mechanism) G with a vocabulary of information granules $L_1, L_2, ..., L_c$ to the membership grade A, we obtain a vector with the following entries:

$$G(A): A \rightarrow [L_1(A)\quad L_2(A)... \quad L_c(A)] \tag{11.10}$$

where $L_1(A), L_2(A), ..., L_c(A)$ are the levels of matching of the numeric membership grade A with the information granules developed during the granulation process for a fixed value of the argument.

While the granular transformation is of a general character, there could be various formal frameworks in which information granules are specified as sets, fuzzy sets, rough sets, and so forth. For instance, if we consider a granulation mechanism G relying on the interval granulation of the space of membership values [0,1], such as L (*low*), M (*medium*), H (*high*) membership, the above granular representation produces a three-dimensional Boolean vector with the entries 0 or 1 as illustrated in Figure 11.10. More formally, the membership degree expressed in terms of the three intervals results in a vector [0 1 0]. For fuzzy sets forming the granules in the unit interval, we obtain a vector of membership grades in the [0,1]-interval.

There is also a two-argument operation ϕ, for which we form the ϕ-consistent granular representation of the fuzzy sets. As noted earlier, the objective is to attain the equality $G(f(A,B)) = f(G(A), G(B))$ with the granular realization of A

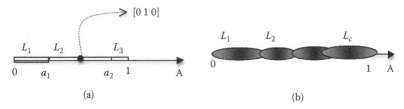

FIGURE 11.10
Interval granulation mechanism applied to membership grade A. (a) $L_1 = low$, $L_2 = medium$, $L_3 = high$, and the resulting three-dimensional Boolean representation vector (with two cutoff points a_1 and a_2). (b) Fuzzy granulation mechanisms involving three fuzzy sets expressed in the unit interval of membership grades.

and B, $G(A)$, $G(B)$, being a subject of the optimization. Formally, we express the optimization task in a continuous (or discrete) format by considering the performance index

$$\int_0^1 \int_0^1 ||G(\phi(A,B)) - \phi(G(A),G(B))|| \, dAdB \qquad (11.11)$$

where the integral (or sum) is taken over all membership grades of A and B taking values in the unit interval. The distance function $||.||$ is typically specified to be the Euclidean one, however the Hamming distance is quite commonly encountered as well. We elaborate on the meaning of Equation (11.10) and the ensuing essence of the minimization of the above performance index.

Let us take a closer look at the performance index and the essence of its minimization. $G(A)$ and $G(B)$ are binary vectors and for the values of A and B confined to L_i and L_k they come with the 1s positioned at the corresponding entries of the vector $A \in L_i$, $G(A) = [0\ 0\ ...0\ 1\ 0\ ...0]$ (the i-th entry set to 1) and $B \in L_k$ $G(B) = [0\ 0...0\ 1\ 0\ ...0]$ (the k-th entry set to 1). $f(G(A), G(B))$ for the values of A and B confined to the information granules as before, is a binary vector $\mathbf{b}_{ik} = [0\ 0\ ...0\ 1\ 0...0]$ in which a certain coordinate, that is, the j-th one is set to 1 and all others are equal to zero. For the values of A and B belonging to some other information granules we have some binary vector $\phi(G(A), G(B))$ with some other coordinate set to 1. Altogether, we can arrange all these indexes for which $\mathbf{b}_{ik}$ assumes 1 in a single vector of indexes **J**. This vector is of dimensionality c^2. For the values of the arguments in L_i and L_k, the expression $f(A, B)$ assumes values in [0,1], not necessarily being confined to the L_j. For the values of A and B in L_i and L_k this means that $G(f(A,B))$ is no longer a binary vector with a single entry equal to 1. Noting that we take a careful look at the integral

$$\int_{A \in L_i} \int_{B \in L_k} ||G(\phi(A,B)) - \phi(G(A),G(B))|| \, dA \, dAB =$$

$$= \int_{A \in L_i} \int_{B \in L_k} ||G(\phi(A,B)) - \mathbf{b}_{ik}|| \, dA \, dAB \qquad (11.12)$$

We can divide the regions of the Cartesian product $L_i \times L_k$ into the subregions Ω_l, $l = 1, 2,..,c^2$ in which $G(\varphi(A, B))$ corresponds to a certain Boolean vector with a single input set to 1. Thus we have $\int_{\Omega_l} \int \| G(\phi(A,B)) - \mathbf{b}_{ik} \| dAdAB$ meaning that the differences produced here quantify the performance of the granular representation of membership functions. By summarizing the above considerations, we note that we have at our disposal the endpoints (bounds) of the intervals in [0,1] and the vector of indexes **J**. As we show in a moment, they become instrumental in the optimization process.

For a given number of information granules (c) and some t-norms and t-conorms specified in advance, the optimization problem can be posed in a formal way as follows:
Minimize

$$\text{MinV}(\mathbf{a}, \mathbf{J}) \,;\, V(\mathbf{a}, \mathbf{J}) = \int_0^1 \int_0^1 \| G(\phi(A, B, \mathbf{a}) - \phi(G(A, \mathbf{a}), G(B, \mathbf{a}))\| dA\, dB$$

subject to constraints

$$0 < a_1 < a_2 < ... < a_c < 1 \tag{11.13}$$

Here, we explicitly show that the granular constructs depend upon the entries of **a** by using the notation $G(f(A, B, \mathbf{a}))$ and $G(A, \mathbf{a})$ as well as $G(B, \mathbf{a})$. The solution to Equation (11.13) arises in the form

$$(\mathbf{a}_{opt}, \mathbf{J}_{opt}) = \arg \text{Min}_{\mathbf{a}, \mathbf{I}} V(\mathbf{a}, \mathbf{J}) \tag{11.14}$$

where the entries of $\mathbf{a}_{opt}$ minimize $V(\mathbf{a}, \mathbf{J})$. The discrete version of the optimization problem (which is more suitable from the practical perspective) is expressed in the form

$$\text{Min } V(\mathbf{a}) \,;\quad V(\mathbf{a}) = \sum_{i=1}^n \sum_{j=1}^n \| G(\phi(A_i, B_j)) - \phi(G(A_i), G(B_j)) \|$$

subject to constraints

$$0 < a_1 < a_2 < ... < a_c < 1$$

$$V(\mathbf{a}) = \sum_{i=1}^n \sum_{j=1}^m \| G(\phi(A_i, B_j, \mathbf{a})) - \phi(G(A_i, \mathbf{a}), G(B_j, \mathbf{a})) \| \tag{11.15}$$

where $A_1, A_2, ... A_n$ and $B_1, B_2, ... B_m$ are the membership grades (membership values) used in the optimization process. We show that the discrete form of the problem is more manageable as it requires an optimization of the vector of the cutoff points of the unit interval, however, the vector of indexes **J** is determined on a basis of the results of logic aggregation of the discrete membership grades (the number of these grades is finite).

For instance, one can consider these grades to be uniformly distributed in the [0,1] interval. In the analogous fashion, we formulate an optimization problem for (ϕ, δ)-consistent granular representation of fuzzy sets. Suppose if f and d are the two-argument operators (such as logic *and* and *or* operators realized by some t-norms and t-conorms), the optimization problem in its continuous version along with its solution reads as follows

$$V(\mathbf{a}, J) = \int_0^1 \int_0^1 \| G(\phi(A, B, \mathbf{a})) - \phi(G(A, \mathbf{a}), G(B, \mathbf{a})) \| dAdB$$

$$+ \int_0^1 \int_0^1 \| G(\delta(A, B, \mathbf{a})) - \delta(G(A, \mathbf{a}), G(B, \mathbf{a})) \| dAdB$$

(11.16)

11.5.2 The Optimization Criterion

The essence of the objective function guiding the optimization problem is to form a vocabulary of information granules so that the results $\phi(G(A), G(B))$ for all membership grades A_k and B_i falling under some fixed elements of the vocabulary are the same granular *meaning*, namely, they invoke the same information granule. Of course, some dispersion can be encountered (meaning that several information granules are invoked) and through an optimization of the information granules, our intent is to minimize it.

To elaborate on this matter, let us consider all possible pairs A_k and B_l, $k = 1, 2, \ldots, n, l = 1, 2, \ldots, m$. Subsequently, we produce the membership grade $\varphi(A_k, B_l)$. Once the granulation process has been realized (see Equation 11.10), the corresponding membership grades A_k and B_l and $\phi(A_k, B_l)$, give rise to a triple of integers corresponding to the elements in the vocabulary of information granules, say i_1, i_2, and i_3. By taking another membership grade, which still produces i_1 and i_2, we obtain another index i_3' (different from i_3). Proceeding with all pairs of membership grades, the results are arranged in the tabular format, namely, a contingency matrix T. The rows and columns of T are labeled by the elements of the vocabulary of information granules used in the realization of *G*. With the (i_1, i_2) coordinate of the matrix T, we associate a vector with the following entries [0 0 … 0 1 0 … 0] where the 1 is located at the *i*-th entry of this vector. The results are successively accumulated in the contingency matrix by updating (incrementing) the content of the vectors associated with the corresponding locations of the table. Thus, the vector with integer coordinates formed at the end of the counting process consists of the counts of occurrence of the elements of the vocabulary used in the granulation process *G*. For the (*i*, *j*)-th coordinate of T we obtain a vector

$$\mathbf{n}_{ij} = [n_{ij}(1)\ n_{ij}(2)\ \ldots\ n_{ij}(c)]^T$$

(11.17)

i, j = 1, 2, ..., c. Denote by k_{ij}^* the largest entry of the vector described by Equation (11.17), that is,

$$k_{ij}^* = \arg\ \max_{k=1, 2, ..., c}\ n_{ij}(k) \qquad (11.18)$$

Refer to Figure 11.11 visualizing a structure of T.

The essence of the ϕ-consistent granular representation of fuzzy sets is to choose the information granules in the granulation scheme G in such a way so that the following performance index being directly linked with Equation (11.17) becomes minimized

$$V(a) = \sum_{i=1}^{c}\ \sum_{j=1}^{c}\ \sum_{\substack{k=1 \\ k \ne k_{ij}^*}}^{c} n_{ij}(k) \qquad (11.19)$$

Formally, we have $Min_a V(a)$ subject to $0 < a_1 < a_2 ... < a_c < 1$. In a nutshell, we select the elements of the vocabulary in such a way that the dispersion of entries of n_{ij} is made as low as possible. The minimization is carried out through the adjustments of the cutoff points (bounds of the intervals) **a**. Note that the optimization is associated with the use of a certain logic operation. This stipulates that the intervals may be different when dealing with a different realization of the logic operator (i.e., a given t-norm).

One can look at the aggregate process of granulation of [0,1] by forming an overall performance index, which combines those used in the formation of the (*or, and*)-consistent granulation of membership functions, namely,

$$V\ (a) = V(and, a) + V(or, a) \qquad (11.20)$$

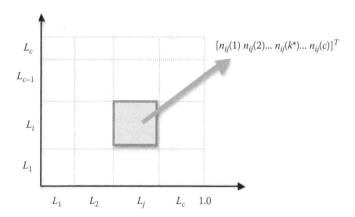

FIGURE 11.11
Optimization of the matrix of occurrences of results of granular processing collected in T; the highest entry marked in **boldface**.

which, as before, is optimized with regard to the coordinates of the vector **a**. This formulation is legitimate considering that a great deal of processing in fuzzy sets is realized by means of *and* and *or* logic operators. We emphasize the existence of the logic operators used in the consistency optimization. By choosing some other t-norm or t-conorm, we arrive at the optimized information granules.

In light of the form of the objective function, which does not depend in an explicit way on the vector of the cutoff points, the use of gradient-based methods where the gradient is computed with respect to the coordinates of (**a**), we have to consider other alternatives such as genetic algorithms (GAs) or particle swarm optimization (PSO). In comparison with GA, PSO is especially attractive given its less significant computing overhead.

11.5.3 Logic-Consistent Granular Representations of Fuzzy Sets: Experiments

We report on a series of experiments in which the f-consistency consists of *and-*, *or-*, and (*and, or*)-consistency with several realizations of t-norms and t-conorms.

In all experiments, PSO was used with the following values of the parameters: the size of population equal to 150, and the number of generations of 200 sets as individuals. These two values were selected as a result of intensive experimentation: it has been observed that after 200 generations there were no further changes (reduction) of the values of the fitness function. The size of the population was found to produce *stable* results meaning that very similar or identical results were reported in successive runs of the PSO. The number of membership grades A_k is equal to 100; these grades are distributed uniformly in the unit interval.

Or-consistency of granular representation. The fitness function is the one expressed by Equation (11.19) with the logic operator realized by several t-conorms. In the series of experiments, we considered a series of cutoff points. The results are contrasted with those obtained when using a uniform distribution of the cutoff points. In all cases, there is a significant improvement quantified in the resulting values of the fitness function as illustrated in Figure 11.12. It is worth noting that in contrast to the relationship V(c) for the uniform distribution of the cutoff points, whose values vary quite significantly with respect to the values of "c," the fitness function for the optimized cutoff points is quite monotonic and with limited changes. All in all, this allocation of the bounds points at the usefulness of optimizing the lengths of the membership intervals rather than adhering to the intervals of the same size.

The granular characterization of the *or* operator for four and seven linguistic terms and the probabilistic sum and the Lukasiewicz *or* operator is presented in the tabular format in Table 11.1 through Table 11.3.

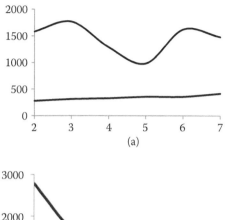

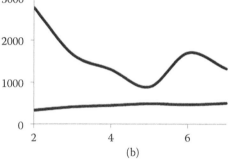

FIGURE 11.12
Values of fitness function V versus number of linguistic terms for selected t-conorms: (a) probabilistic sum, (b) Lukasiewicz *or*. The upper curves in the figures pertain to the uniform distribution of the cutoff points.

TABLE 11.1

Tabular Description of the *Or* Operation*

Granular terms	L_1	L_2	L_3	L_4
L_1	L_1	L_2	L_3	L_4
L_2	L_2	L_3	L_4	L_4
L_3	L_3	L_4	L_4	L_4
L_4	L_4	L_4	L_4	L_4

*T-conorm: probabilistic sum, 4 linguistic terms.

TABLE 11.2

Tabular Description of the *Or* Operation*

Granular terms	L_1	L_2	L_3	L_4
L_1	L_1	L_2	L_3	L_4
L_2	L_2	L_3	L_4	L_4
L_3	L_3	L_4	L_4	L_4
L_4	L_4	L_4	L_4	L_4

*T-conorm: Lukasiewicz *or*, 4 linguistic terms.

TABLE 11.3

Tabular Description of the *Or* Operation*

	L_1	L_2	L_3	L_4	L_5	L_6	L_7
L_1	L_1	L_2	L_3	L_4	L_5	L_6	L_7
L_2	L_2	L_3	L_4	L_5	L_6	L_7	L_7
L_3	L_3	L_4	L_5	L_6	L_7	L_7	L_7
L_4	L_4	L_5	L_6	L_7	L_7	L_7	L_7
L_5	L_5	L_6	L_7	L_7	L_7	L_7	L_7
L_6	L_6	L_7	L_7	L_7	L_7	L_7	L_7
L_7	L_7	L_7	L_7	L_7	L_7	L_7	L_7

**t*-conorm: Lukasiewicz *or*, c = 7.

For the case of c = 4, the table can be expressed in the following form:

$$A \ or \ B =$$

$$\begin{cases} L_1 & \text{if } A \in L_1, \ B \in L_1 \\ L_2 & \text{if } A \in L_2, B \in L_1 \ \text{or} \ A \in L_1, B \in L_2 \\ L_3 & \text{if } A \in L_1, B \in L_3 \ \text{or} \ A \in L_3, B \in L_1, \ A \in L_2, B \in L_2 \\ L_4 & \text{otherwise} \end{cases}$$

Two interesting observations are to be made here. First, the information granules (labels) need to be optimized and the optimized version produces tangible advantages in terms of the consistency of the granular realization of the logic operator. Second, the tabular representation makes the realization of the logic operator more explicit. We might have been anticipating the general character of the *or* operator (where, in general, the larger index of the arguments of the operator is produced), however, the table offers the details as to the nature of the mapping. We also observe a certain elevation effect and interaction invoked. If one argument is L_3, sufficiently high granular values of the second one produce L_7 (see Table 11.3).

11.6 Modes of Processing with Granular Characterization of Fuzzy Sets

The granular representation of fuzzy sets gives rise to two modes of processing in which the granularity of information manifests itself in various ways. The first one follows the general scheme coming in the form

Fuzzy sets A, B, C ...→ granular representation *G*(A), *G*(B), *G*(C)...→
processing at the level of information granules

The processing at the level of information granules invokes the use of the tabular representation of operations on fuzzy sets. The computing is realized at the granular (symbolic, nonnumeric) level of information granules meaning that the amount of computing is quite limited.

The second mode of processing adheres to the processing of fuzzy sets at the level of numeric membership values while the final result is represented as a granular construct. In other words, we have

$$\text{Fuzzy sets A, B, C}... \rightarrow \text{processing} \rightarrow \text{fuzzy set } \Omega \rightarrow \text{granular}$$
$$\text{representation } G(\Omega)$$

The granular representation of Ω results in a collection of pairs (Ω_i, L_i) where the information granule L_i occupies the corresponding region of the universe of discourse (Ω_i).

Figure 11.13 highlights these two modes of processing and distinguishes between the numeric and granular processing of membership grades, which result in two well-articulated levels of hierarchy.

We provide two illustrative examples, which highlight the nature of the granular representation of fuzzy sets and relate to the logic operators forming the crux of the underlying computing. In both cases, we are concerned with the (*or, and*)-consistent granular representation of fuzzy sets.

Compositions of fuzzy sets and fuzzy relations. The main categories of composition operators rely on the usage of logic operators, say max-min, min-max, max-t, min-s, and s-t compositions. For the tabular representations of the corresponding logic operators, the calculations of the composition are based on the use of the tables so this processing becomes of particular interest when dealing with implementations in the presence of limited computing resources. For instance, let us consider the product realization of the logic *and* and *or* implemented by

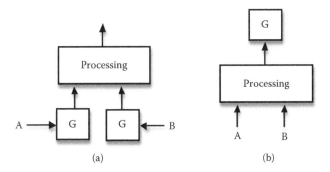

(a)   (b)

FIGURE 11.13
Two modes of processing involving information granularity.

the probabilistic sum, the composition of A and R, A ∘ R, with the entries

$$A = [0.82\ 1.00\ 0.45\ 0.12\ 0.78\ 0.65]\ \text{and}\ R = \begin{bmatrix} 0.79 & 0.43 \\ 0.50 & 0.27 \\ 0.11 & 1.00 \\ 0.32 & 0.56 \end{bmatrix}$$

The granular representation is (*or, and*)-consistent and with c = 6 information granules of fuzzy sets; refer to the tabular representation of logic operators. This gives rise to the granular representation of A and R to come in the form

$$A = [L_3, L_4, L_6, L_1]\quad R = \begin{bmatrix} L_6 & L_3 \\ L_3 & L_2 \\ L_1 & L_6 \\ L_2 & L_4 \end{bmatrix}$$

Then the *or-and* composition operator produces the granular result A ∘ R = $[L_4, L_6]$.

11.7 Conclusions

Decision making is inherently a human-driven pursuit. The challenge of modeling this pursuit and arriving at the models that properly capture the linguistic fabric of the terms used by humans is to translate the linguistic (qualitative) terms into a certain operational framework of information granules. We showed that the translation problems are expressed as some optimization problems. We showed that information granularity plays a pivotal role when building consensus in group decision-making problems where the individual points of view are put together in the granular rather than numeric format.

In the context of linguistic information and their formal granular models, it is worth making a clear distinction between vertical and horizontal granulation of information and the associated levels of information granularity. With regard to the vertical granulation, we are interested in the formation of a collection of information granules describing membership allocation to a given information granule. The examples of this granulation were presented in this chapter. In contrast, the horizontal granulation deals with the formation and distribution of information granules across the universe of discourse; a great many constructs falling under this rubric have been presented in the previous chapters.

References

Aliev, R.A., B. Fazlollahi, and R.R. Aliev. 2004. *Soft Computing and Its Applications in Business and Economics*. Heidelberg: Springer-Verlag.

Apostolou, B. and J.M. Hassell. 2002. Note on consistency ratio: A reply. *Mathematical and Computer Modeling*, 35, 1081–1083.

Çakır, O. 2008. On the order of the preference intensities in fuzzy AHP. *Computers & Industrial Engineering*, 54, 993–1005.

Chu, P. and J.K.H. Liu. 2002. Note on consistency ratio. *Mathematical and Computer Modeling*, 35, 1077–1080.

Dong, Y., G. Zhang, W.C. Hong, and Y. Xu. 2010. Consensus models for AHP group decision making under row geometric mean prioritization method. *Decision Support Systems*, 49, 281–289.

Frei, F.X. and P.T. Harker. 1999. Measuring aggregate process performance using AHP. *European Journal of Operational Research*, 116, 436–442.

Herrera-Viedma, E., S. Alonso, F. Chiclana, and F. Herrera. 2007. A consensus model for group decision making with incomplete fuzzy preference relations. *IEEE Transactions on Fuzzy Systems*, 15, 863–877.

Herrera, F., E. Herrera-Viedma, and J.L. Verdegay. 1997. A rational consensus model in group decision making using linguistic assessments. *Fuzzy Sets & Systems*, 88, 1, 31–49.

Herrera-Viedma, E., L. Martinez, F. Mata, and F. Chiclana. 2005. A consensus support system model for group decision-making problems with multigranular linguistic preference relations. *IEEE Transactions on Fuzzy Systems*, 13, 644–658.

Jeonghwan, J., L. Rothrock, P.L. McDermott, and M. Barnes. 2010. Using the analytic hierarchy process to examine judgement consistency in a complex multiattribute task. *IEEE Transactions on Systems, Man and Cybernetics, Part A: Systems and Humans*, 40, 1105–1115.

Kangmao, W., C.K. Wang, and H. Chun. 2005. Analytic hierarchy process with fuzzy scoring in evaluating multidisciplinary R&D projects in China. *IEEE Transactions on Engineering Management*, 52, 119–129.

Korpela, J. and M. Tuominen. 1996. Benchmarking logistics performance with an application of the analytic hierarchy process. *IEEE Transactions on Engineering Management*, 43, 323–333.

McCauley-Bell, P. and A.B. Badiru. 1996. Fuzzy modeling and analytic hierarchy processing to quantify risk levels associated with occupational injuries—Part I: The development of fuzzy–linguistic risk levels. *IEEE Transactions on Fuzzy Systems*, 4, 124–131.

Mizumoto, M. and K. Tanaka. 1981. Fuzzy sets of type-2 under algebraic product and algebraic sum. *Fuzzy Sets and Systems*, 5, 277–290.

Mustafa, M.A. and J.F. Al-Bahar. 1991. Project risk assessment using the analytic hierarchy process. *IEEE Transactions on Engineering Management*, 38, 46–52.

Pedrycz, W. 2009. Statistically grounded logic operators in fuzzy sets. *European Journal of Operational Research*, 193, 2, 520–529.

Saaty, T.L. 1983. Introduction to a modeling of social decision process. *Mathematics and Computers in Simulation*, 25, 105–107.

Saaty, T.L. 1987. How to handle dependence with the analytic hierarchy process. *Mathematical Modelling*, 9, 369–376.

Saaty, T.L. and G. Hu. 1998. Ranking by eigenvector versus other methods in the analytic hierarchy process. *Applied Mathematics Letters*, 11, 121–125.

Saaty, T.L. and M. Ozdemir. 2003. Negative priorities in the analytic hierarchy process. *Mathematical and Computer Modelling*, 37, 1063–1075.

Tung, S.L. and S.L. Tang. 1998. A comparison of the Saaty's AHP and modified AHP for right and left eigenvectors inconsistency. *European Journal of Operational Research*, 106, 123–128.

Van Laarhoven, P.J.M. and W. Pedrycz. 1983. A fuzzy extension of Saaty's priority theory. *Fuzzy Sets & Systems*, 11, 229–241.

Wang, T.C. and Y.H. Chen. 2008. Applying fuzzy linguistic preference relations to the improvement of consistency of fuzzy AHP. *Information Sciences*, 178, 3755–3765.

Zadeh, L.A. 1975. The concept of a linguistic variable and its application to approximate reasoning. *Information Sciences*, 8, 199–249.

Zadeh, L.A. 2011. A note on Z-numbers. *Information Sciences*, 181, 2923–2932.

Index